Konrad Deetz

Assessing the Aerosol Impact on Southern West African Clouds and Atmospheric Dynamics

Wissenschaftliche Berichte des Instituts für Meteorologie und Klimaforschung des Karlsruher Instituts für Technologie (KIT)
Band 75

Herausgeber: Prof. Dr. Ch. Kottmeier

Institut für Meteorologie und Klimaforschung
am Karlsruher Institut für Technologie (KIT)
Kaiserstr. 12, 76128 Karlsruhe

Eine Übersicht aller bisher in dieser Schriftenreihe
erschienenen Bände finden Sie am Ende des Buches.

Assessing the Aerosol Impact on Southern West African Clouds and Atmospheric Dynamics

by
Konrad Deetz

KIT Scientific Publishing

Dissertation, Karlsruher Institut für Technologie
KIT-Fakultät für Physik

Tag der mündlichen Prüfung: 3. November 2017
Referenten: Prof. Dr. Ch. Kottmeier, Prof. Dr. P. Knippertz

Impressum

KIT Scientific Publishing

Karlsruher Institut für Technologie (KIT)
KIT Scientific Publishing
Straße am Forum 2
D-76131 Karlsruhe

KIT Scientific Publishing is a registered trademark
of Karlsruhe Institute of Technology.
Reprint using the book cover is not allowed.

www.ksp.kit.edu

Print on Demand 2018 – Gedruckt auf FSC-zertifiziertem Papier

ISSN 0179-5619
ISBN 978-3-7315-0744-4
DOI 10.5445/KSP/1000077925

Assessing the Aerosol Impact on Southern West African Clouds and Atmospheric Dynamics

Zur Erlangung des akademischen Grades eines

DOKTORS DER NATURWISSENSCHAFTEN

von der Fakultät für Physik des

Karlsruher Instituts für Technologie (KIT)

genehmigte

DISSERTATION

von

M.Sc. (Meteorologie) Konrad Deetz

aus Zehdenick

Tag der mündlichen Prüfung:	3. November 2017
Referent:	Prof. Dr. Ch. Kottmeier
Korreferent:	Prof. Dr. P. Knippertz

Abstract

Southern West Africa undergoes rapid and significant socioeconomic changes related to a massive increase in air pollution. The impact of the atmospheric pollutants on weather and climate in this region is virtually unexplored. In this study, the regional-scale model framework COSMO-ART is applied to southern West Africa during the West African summer monsoon period to quantify the atmospheric composition and to assess the aerosol direct and indirect effect on the meteorological conditions. It thus addresses one of the main topics of the project Dynamics-Aerosol-Chemistry-Cloud Interactions in West Africa (DACCIWA).

The model application is accompanied with significant further developments of the emission parameterizations regarding mineral dust and the biogenic compounds. Furthermore, a parameterization for trace gas emissions from gas flaring of the oil industry was developed to reproduce the specific pollution conditions of the research area.

COSMO-ART is for the first time used as an operational atmospheric composition forecast model for a three-month period to support the aircraft missions of the DACCIWA measurement campaign. The forecast results allowed an extensive evaluation of the meteorological, trace gas and aerosol state of southern West Africa. Especially good agreement is achieved in the representation of the nocturnal low-level stratus, a cloud feature that frequently evolves during night and persists to the early afternoon. The modeled spatiotemporal distribution of atmospheric pollutants is in accordance to observations and shows small day-to-day variability with the formation of distinct pollution plumes that emerge from the coastal cities and are advected northeastwards with the monsoon flow. However, COSMO-ART shows a tendency to overestimate the concentration of atmospheric pollutants from anthropogenic origin in southern West Africa.

Within a case study focusing on 2-3 July 2016, the meteorological responses to changes in the aerosol amount are assessed. The results revealed that a considerable fraction of the aerosol in the moist monsoon layer is water, significantly altering the AOD and therefore the aerosol direct effect. About 58 % of the ALWC is related to the mixed accumulation mode. As indicated in previous studies, a coastal front was observed that propagates inland in the evening. For a differentiation from the land-sea breeze, the term *Evening Monsoon Flow Enhancement* (EMFE) was introduced. Strongest responses of the EMFE are clearly related to the aerosol direct effect. A reduction in the aerosol amount leads to two feedback chains comprising a spatial shift of the EMFE front of up to 25 km and a temporal shift of the stratus-to-cumulus transition of about 1 hour. The aerosol indirect effect shows the previously known effects of an increase in cloud droplet number concentrations, effective radii and suppression of precipitation with increased aerosol but no significant microphysical differences are detected between the nocturnal low-level stratus in different regions or inside and outside of city pollution plumes. The inland aerosol amount is generally high, likely decreasing the susceptibility of clouds to changes in aerosol.

More research is necessary regarding the aerosol water uptake in SWA and their influence on the radiation budget also on longer time scales. Also further approaches to disentangle the aerosol indirect effect from the total aerosol response are necessary. The results of this study underscore the need for more reliable anthropogenic emission datasets and a continuous monitoring of trace gases and aerosols in the coastal cities and their surrounding areas.

Contents

1 Introduction

Aerosol is the general term for all liquid and solid particles suspended in the atmosphere. Atmospheric aerosol particles are highly relevant in terms of weather, climate and human health, since they modify the formation of clouds and precipitation, alter the global radiation budget by scattering and absorption and can have adverse effects on the human respiratory system. With regard to globally accelerated industrialization and urbanization, linked with increased emissions of anthropogenic pollutants in particular in developing and newly industrializing countries, aerosol research is of great importance in atmospheric science but also with respect to socioeconomic questions.

The project *Dynamics-Aerosol-Chemistry-Cloud Interactions in West Africa* (DACCIWA; Knippertz et al., 2015a, www.dacciwa.eu) is a consortium of 16 research institutes from the six countries Germany, United Kingdom, Switzerland, France, Ghana and Nigeria, focusing on the densely populated coastal area of southern West Africa (SWA). In contrast to the *African Monsoon Multidisciplinary Analysis* (AMMA, Redelsperger et al., 2006), as the most comprehensive research program focusing on West Africa to date, DACCIWA moves the focus from the Sahelian region to the coastal region at the Gulf of Guinea. UNO (2015b) indicates that more than the half of the global population growth between now and 2050 will be related to Africa. For Nigeria, that has a population of 182 million in 2015 (rank 7), a population increase to 399 million (rank 3) is expected. Accompanied with these projections, Liousse et al. (2014) showed that the African anthropogenic emissions will significantly increase from 2005 to 2030 if no emission regulations were implemented.

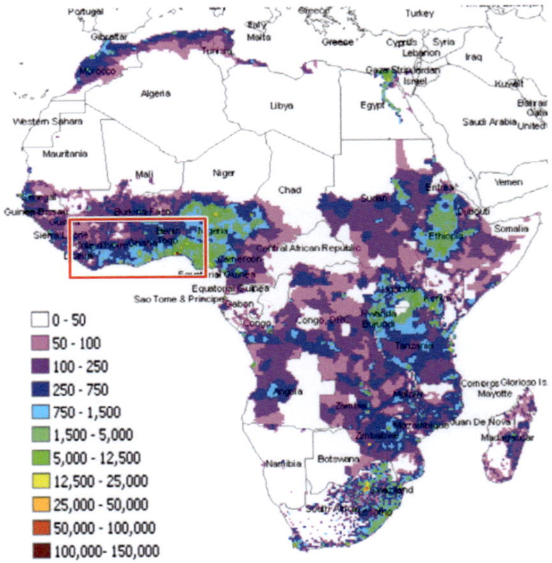

Figure 1.1: Projection of the spatial distribution of organic carbon emissions from anthropogenic combustion processes in Africa (tons of carbon yr^{-1}) for 2030 in a business-as-usual scenario (Liousse et al., 2014, adapted). The red rectangle denotes the location of SWA.

In the business-as-usual scenario an increase in emissions of carbon monoxide (CO), nitrogen monoxide and dioxide (NO_x) and sulfur dioxide (SO_2) by factors of 3.2, 5.6 and 6.5, respectively, is projected from 2005 to 2030. Figure 1.1 shows the projected organic carbon emissions for Africa in 2030. Furthermore, UNO (2015a) exhibits that simutaneously the urbanization will increase. For Nigeria the population fraction living in urban areas is estimated or projected with: 47% (2014), 54.1% (2030) and 62.7% (2050). By taking also into account the land use changes, the global climate change and their impacts on the WAM, the region is and will be subject to substantial alteration. On the other hand for this area the data basis in terms of meteorological characteristics and air pollution, in particular the cloud microphysical properties and aerosol loading, is poor. To overcome these shortcomings, a DACCIWA-related comprehensive field campaign took place in June-July 2016 including extensive ground-based measurements (Kalthoff et al., 2017) and airborne measurements (Flamant et al., 2017).

DACCIWA comprises expertise in modeling and observation including weather prediction and climate science, air pollution, cloud microphysics and radiation and deploys this expertise to shed light on the area of interest. Within the work packages Boundary Layer Dynamics, Air pollution and Health, Atmospheric Chemistry, Cloud-Aerosol Interactions, Radiative Processes, Precipitative Processes and Monsoon Processes, the interrelation including the emissions of atmospheric pollutants, the aerosol influence on clouds, precipitation, radiation and atmospheric dynamics is addressed. Furthermore, SWA is frequently covered by a low-level stratus deck (e.g. Schrage and Fink, 2012; Schuster et al., 2013; Adler et al., 2017) for which only hypotheses of formation mechanisms exists. Especially the interaction between the stratus and air pollution is not quantified up to now. DACCIWA is introduced in detail in Knippertz et al. (2015a). In Knippertz et al. (2015b) the key research questions behind the project and the link to future perspectives beyond DACCIWA are elaborated.

In the framework of DACCIWA this study focuses on the quantification of the aerosol direct effect (ADE) and the aerosol indirect effect (AIE) on warm stratiform and cumuliform clouds over SWA. The diversity in sources of atmospheric aerosols together with the open questions with respect to their feedbacks on the atmospheric dynamics and the low-level stratus and in combination with the extensive datasets obtained during the DACCIWA campaign, makes SWA an ideal region for a modeling study of ADE and AIE. Within this work, the regional-scale model framework COSMO-ART (Vogel et al., 2009) is applied to address the following scientific questions:

- *Is it feasible to run COSMO-ART in an operational forecast mode for SWA?*
- *Which typical spatiotemporal patterns of the atmospheric composition can be identified?*

- *Can robust ADE and AIE on SWA clouds and atmospheric dynamics be detected?*
- *In which respect are low-level clouds affected by changes in the aerosol amount?*

This study is structured as follows: In Section 2 the research domain SWA is introduced in more detail and the current state of research in meteorological, atmospheric composition and aerosol-atmosphere interaction perspective is presented. Section 3 describes the model COSMO-ART. The model description explicitly includes work that is realized in this study, especially relating to the further or new development of atmospheric pollution emission parameterizations. These contributions will be highlighted in the respective area. The datasets that are used for the COSMO-ART realizations and the model evaluation are depicted in Section 4. The two subsequent sections address the research questions presented above. During the DACCIWA campaign, operational forecasts with regularly updated actual boundary conditions were conducted with COSMO-ART as a part of this study. Section 5 describes their realization together with a detailed analysis of the forecast performance by using data from the DACCIWA campaign and other sources. The aerosol feedback analysis in Section 6 focuses in detail on the quantification of ADE and AIE in SWA. The study concludes with a summary and evaluation of the findings with regard to the posed scientific questions (Sect. 7).

2 Scientific Background

2.1 The Research Domain Southern West Africa (SWA)

This study focuses on SWA with the key domain 5°N-10°N and 8°W-8°E, including the countries Ivory Coast, Ghana, Togo, Benin and Nigeria and the Gulf of Guinea (see Fig. 2.1). The DACCIWA ground-based measurement campaign (Sect. 4.2) includes three so-called supersites. They are located at Kumasi (Ghana), Savè (Benin) and Ile-Ife (Nigeria), as denoted in red in

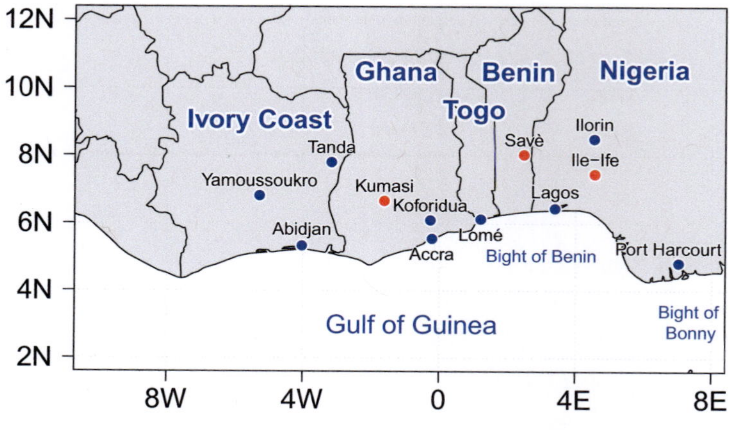

Figure 2.1: Map of the SWA research domain. The red dots mark the three supersites Kumasi, Savè and Ile-Ife.

Figure 2.1. The other cities labeled in Figure 2.1 (denoted in blue) are either big cities, highly relevant as sources of anthropogenic emissions, or are related to other observational sites used in this study.
Table 2.1 summarizes the five cities with the highest population in SWA (UNO, 2016), each greater than 2.3 million. These cities are, aside from Kumasi, exclusively located at the coast. Lagos in southwestern Nigeria has the status of a megacity (population greater than 10 million). To allow for comparison, the German capital Berlin is added to the table.

Table 2.1: The five SWA cities with the highest population in 2016 (UNO, 2016). The German capital Berlin is added for comparison and the reference area denotes the area that is considered for the population quantification. *City proper* encompasses the area within the city limits, *Urban Agglomeration* is the continuous urbanized area and the *Metropolitan Area* also includes less-populated surrounding territories.

City	Reference area	Population (m)
Lagos	Urban Agglomeration	13.7
Abidjan	Urban Agglomeration	5.0
Kumasi	Metropolitan Area	2.7
Port Harcourt	Urban Agglomeration	2.5
Accra	Metropolitan Area	2.3
(Berlin	City Proper	3.6)

In addition, Figure 2.2 exhibits the topography of SWA. In general, the domain shows a zonally relatively uniform gradual increase in elevation from near sea level in the South to values of about 500 m above sea level (ASL) in the North. Special topographic features are located in the center of the SWA domain (see Fig. 2.2). The Mampong Range extends from the Southeast of Ghana to the Northeast of Ivory Coast. In northeasterly direction the Atakora Mountains extends over Togo and northern Benin.

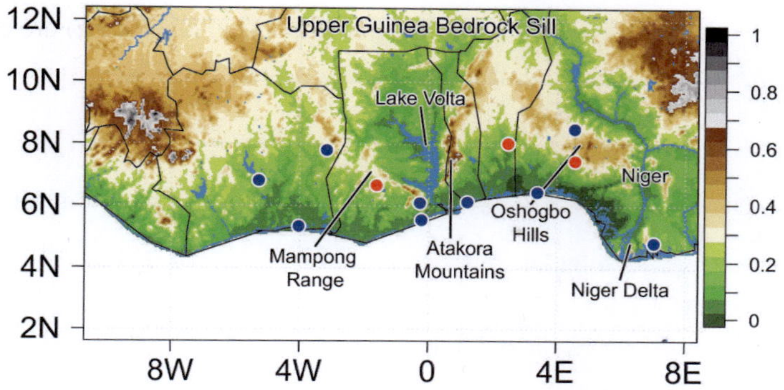

Figure 2.2: Map of the SWA research domain regarding the topography (km ASL). The red dots mark the three supersites Kumasi, Savè and Ile-Ife (compare Fig. 2.1).

The Mampong Range and the Atakora Mountains encompass the Volta basin with Lake Volta and the plain northward. Over southwestern Nigeria and northern Benin the Oshogbo Hills extend in northwesterly direction. The Niger Delta, the mouth of the Niger river, is located in the southeastern part of Nigeria. SWA is characterized by a Tropical-Monsoon climate according to the Köppen-Geiger climate classification (Peel et al., 2007). In the southern part of Ivory Coast, the southwestern part of Ghana and the southernmost part of Nigeria evergreen tropical forest is dominating. To the North the vegetation is changing to savanna and grasslands.

2.2 Meteorological Characteristics of SWA

This section encompasses an introduction to the meteorological characteristics of SWA, starting with the WAM as the principle meteorological driver, subsequently steering the focus to the implications on the SWA region and concluding with the conditions during the DACCIWA campaign in 2016 via a classification of the campaign period in meteorological phases (Knippertz et al., 2017) and first outcomes from the meteorological observations obtained at the SWA supersites (Kalthoff et al., 2017).

2.2.1 The West African Monsoon (WAM)

The meteorological situation in SWA is governed by the WAM. The WAM is characterized by a strong seasonal cycle with moist, cool southwesterly (monsoon) winds during boreal summer and dry (Harmattan) winds during boreal winter. The boundary between the moist southwesterly and the dry northeasterly flow is the intertropical discontinuity (ITD). West Africa is characterized by a marked south-north gradient in the conditions at the surface, as described in Section 2.1. This includes the topography and the surface itself, from the Gulf of Guinea, to tropical forests, savanna and the Sahara Desert.
From November to February the ITD is located south of 10°N (approximately between 2°N and 10 °N) (Hayward and Oguntoyinbo, 1987). Therefore SWA can be affected by the dry Harmattan winds transporting mineral dust from the Sahara (dry season). In this case, clouds and precipitation are restricted to the Guinea coast. In early May, with the northward movement of the ITD, the southwesterly monsoon winds start to penetrate deeper inland. Sultan and Janicot (2003) define the monsoon pre-onset by the arrival of the ITD at 15°N while the intertropical convergence zone (ITCZ) is centered at 5°N and the monsoon onset when the ITCZ position shifts from 5°N to 10°N. With the monsoon onset, the rainfall maximum shifts from the coast to the Sahel. The peak of the wet season is reached in July and August. For Ghana, Amekudzi et al. (2015) analyzed a 42-year precipitation climatology, identifying the rainfall onset between March and May with a length of the rainy season of 240 ± 7 days in the coastal area and 140 ± 5 days in northern Ghana. Within the summer monsoon, SWA is characterized by strong southwesterly winds driven by the pressure gradient between the cool sea surface temperature (SST) of the eastern equatorial Atlantic Ocean and the Saharan heat low (SHL). Figure

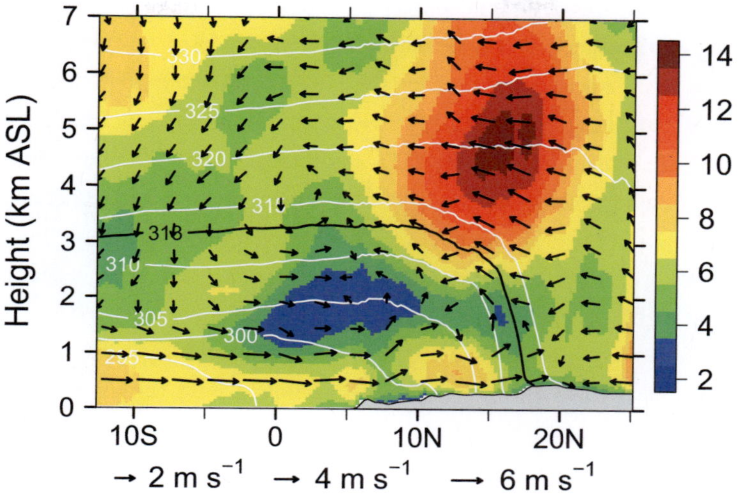

Figure 2.3: Meridional vertical transect (km ASL) at the Greenwich meridian illustrating the WAM circulation via the horizontal wind speed (m s^{-1}, color shading) and atmospheric circulation (meridional and vertical component, each normalized with their maximum to emphasize the vertical transport, arrows). The bottom scale assigns the non-normalized wind speed values (approximated) to the length of the arrows. Furthermore, the isolines of virtual potential temperature (K, white contours) are shown. The black isentrope (θ_v=313 K) is used as an indicator of the monsoon layer top and the lower limit of the SAL following Parker et al. (2005a) and Parker et al. (2005b). The figure is based on the 21 UTC temporal average of the COSMO-ART forecasts of the *Post-onset phase* of the monsoon in 2016 between 22 June to 20 July. The gray shading denotes the topography.

2.3 shows a meridional vertical transect along the Greenwich meridian for the temporal average of the post-onset monsoon phase in 2016 (29 d-average using the time step 21 UTC from 22 June to 20 July 2016) to illustrate the atmospheric circulation within the WAM. The figure is based on COSMO-ART forecasts realized in the framework of this study introduced in Section 3.1.5. The DACCIWA research domain is located between 5°N and 10°N. The large meridional low-level temperature and pressure gradient between the eastern equatorial Atlantic Ocean (south of 5°N), in particular the Atlantic Cold

Tongue (Caniaux et al., 2011), and the SHL (between 18°N and 30°N) leads to a strong southwesterly monsoon flow in the lowest 1-3 km, reaching 20°N, that is linked to enhanced baroclinicity (see Fig. 2.3). Around 20°N the ITD can be found, indicating the boundary between the moonsoon flow from the South and the SHL in the North. At the ITD, a northerly return flow above the actual mosoon flow is visible that can lead to enhanced mineral dust concentrations in higher altitudes. With the daytime development of the planetary boundary layer (PBL), dry air from the return flow can be mixed down that is denoted by Parker et al. (2005a) as a diurnal *flushing* of the monsoon layer.

Furthermore, high pressure above the SHL induces northerly winds, that extend the Saharan Air Layer (SAL) southwards, linked with further mineral dust burden in altitudes between 3 and 7 km. According to Parker et al. (2005b), the boundary between the monsoon layer and the SAL can be estimated via the location of the 313 K isoline of virtual potential temperature (see black line in Fig. 2.3).

The southern edge of the SAL is marked by a strong latitudinal band of easterly winds, the African Easterly Jet (AEJ), located between 10-20°N and at 3-7 km ASL with the jet core at 15°N and 4-5 km ASL for the situation presented in Figure 2.3. The AEJ results from the strong baroclinicity across the ITD together with the contrasts in moist convection south of the ITD and dry convection north of the ITD (Lafore et al., 2011). According to the thermal wind relation, easterly shear above the mosoon layer and therefore easterly winds develop (Cook, 1999). In Figure 2.3, the AEJ axis is perpendicular to the wind speed maximum at 15°N and 4-5 km ASL, with a flow pointing into the drawing plane.

The major synoptic weather systems of the WAM are the African Easterly Waves (AEW) that are generated by the barotropic and baroclinic instability at the AEJ (Lafore et al., 2011) and lead to a meandering structure of the AEJ. Furthermore, the vertical wind shear between the AEJ and the monsoon layer favors the development of Mesoscale Convective Systems (MCS) (Mathon et al., 2002a) that follow the AEW and are highly relevant for the precipitation in the Sahelian region (Mathon et al., 2002b). A detailed description of the WAM circulation can be found e.g. in Chang et al. (2011) and Laing and Evans (2011).

2.2.2 Meteorological Conditions in SWA

By focusing on SWA, Hill et al. (2016) derived multisatellite climatologies of meteorological parameters for June-July. They showed that the mean cloud cover of 68-94% is higher than the global average of 68%. The average outgoing shortwave (longwave) radiation is 145 (230) W m^{-2}. SWA is characterized by an average amount of daily precipitation of 5.87 mm. Linked to the WAM, nocturnal low-level stratus (NLLS hereafter) and stratocumulus decks form frequently over SWA (e.g. Schrage and Fink, 2012; Schuster et al., 2013; van der Linden et al., 2015; Adler et al., 2017). van der Linden et al. (2015) derived NLLS climatologies from remote sensing observations for the time period 2006-2011 indicating that approximately 800.000 km^2 of SWA are affected by NLLS. Schrage and Fink (2012) and Schuster et al. (2013) used an observational and modeling approach, respectively, to assess the characteristics of the NLLS by focusing on the AMMA period 2006. Schrage and Fink (2012) have identified a clear seasonal cycle of NLLS occurrence for 2006 with a maximum in August. The cloudiest area is located north of the Guinean coast and south of 9°N. The maximum (minimum) NLLS extension is reached around 9 UTC (18 UTC). After 10 UTC the NLLS deck breaks up and transforms to cumuliform clouds (Schrage and Fink, 2012; van der Linden et al., 2015).

Schrage and Fink (2012) identified the nocturnal low-level jet (NLLJ hereafter) as the pivotal mechanism for the NLLS formation by producing vertical upward mixing of moisture from the near-surface layer favoring cloud formation. A NLLJ prevails if the wind speed maximum is more than 2 m s^{-1} faster than wind speeds above within the lowest 1500 m above ground level (AGL)

(Stull, 1988). It reaches maximum velocities on an order of magnitude of 10-20 m s^{-1} at a height of 100-300 m AGL and can reach spatial extensions of hundreds or thousands of kilometers. The NLLJ occurs in many regions of the world. Two prominent examples are the Koorin nocturnal jet in Australia (Brook, 1985) and the Great Plains nocturnal jet in the United States (e.g. Song et al., 2005). Several causes for the formation of low-level jets exist. The most important mechanism for the NLLJ formation are inertial oscillations (Blackadar, 1957; van de Wiel, 2010). During daytime, winds in the mixing layer are subgeostrophic and cross the isobars toward low pressure due to friction. After sunset, the turbulence decreases and longwave cooling of the surface leads to a near-surface inversion. The inversion layer and the layer above decouple, which is related to a significant reduction in friction of the latter. This reduction in friction accelerates the wind in direction of the pressure gradient and simultaneously the Coriolis force induce an inertial oscillation. Similar to the description of an harmonic oscillator (with the Coriolis force as restoring force) the wind vectors describe a circle with the inertial period $\frac{2\pi}{f}$, where f is the coriolis parameter. Therefore the wind shows an oscillation in direction and velocity. After sunrise, the NLLJ is destroyed via the reformation of the mixing layer.

With the Coriolis force, the SWA nocturnal low-level winds receive an amplified westerly component. This leads in the morning also to a stronger near-surface zonal component by downward mixing of momentum from the height of the jet maximum (Parker et al., 2005a). The same mechanism is also responsible for the near-surface wind speed maximum in the morning. Schuster et al. (2013) identified the maximum of the SWA NLLJ (jet axis) at similar heights as the NLLS within a model study (grid mesh size of 3 km). Furthermore, the study reveals that the NLLJ is stronger in cloudy nights than in clear nights. NLLS forms after sunset along the Guinea coast and penetrates inland subsequently, due to advection and the distribution of turbulent conditions.

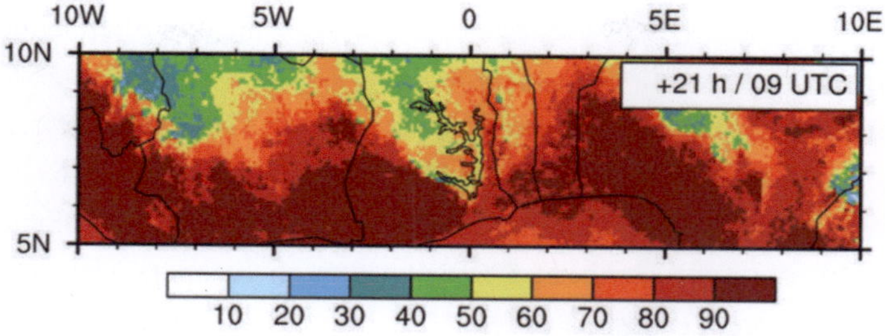

Figure 2.4: Spatial distribution of NLLS coverage (%) at 9 UTC, based on a model climatology for the period June-September 2006 over SWA (Schuster et al., 2013, adapted).

The NLLS cloud base is between 200-400 m AGL (Schrage and Fink, 2012; Schuster et al., 2013) increasing in the course of the morning (Schrage and Fink, 2012). At the NLLS maximum around 9 UTC, the stratus deck on average covers the southern half of Ivory Coast, the southwestern part of Ghana (south of the Mampong Range) and most areas east of the Atakora Mountains (Fig. 2.4). The shortwave radiation difference between less cloudy and cloudy regions is quantified with -35 W m^{-2} and the NLLS is not related to significant precipitation amounts (only light drizzle). Schuster et al. (2013) described the mechanisms that form NLLS separately for the coastal and inland area, whereupon the advection of cool and dry air (relative to inland) is relevant for both areas. In the coastal area, the vertical upward mixing due to the turbulence below the NLLJ is stronger compared to the inland conditions. The vertical mixing leads to an upward transport of latent heat (warming of the cloud layer) and a downward transport of sensible heat (cooling of the cloud layer). Together with the radiative cooling at the cloud top, a net cooling of the cloud layer is established that maintains the clouds. The higher the surface radiative cooling, the stronger the decoupling between surface and NLLJ, the weaker the vertical mixing and the less effective the cloud formation. After the formation of clouds, the surface cooling is reduced, making the vertical mixing more effective and maintaining the clouds (NLLS mechanism 1). Regarding the inland conditions, Schuster et al. (2013) indicated a minor relevance of the vertical mixing since the turbulence is reduced compared to the coastal area. Instead a combination of cool air advection (as for the coastal area), vertical mixing via the NLLJ and topographically forced lifting forms the cloud and radiative cooling supports their maintenance (NLLS mechanism 2).

The model study of Adler et al. (2017) (grid mesh size of 0.5 km) described the NLLS temporal evolution over southern Benin within three phases: In the first phase (18-22 UTC), the stable stratified nocturnal PBL is influenced by cooling and moistening with the passage of a northward propagating front (described below). In the second phase (22-3 UTC), clouds are generated in mountainous areas due to topographically forced lifting and one hour later in non-mountainous areas due to further cooling and moistening and in crests of gravity waves. The third phase (3-8 UTC) is characterized by an increase in the water content and spatial coverage of the clouds. Adler et al. (2017) introduce two additional NLLS formation mechanisms with regard to Schuster et al. (2013). The first is the cloud formation via vertical cold air advection in gravity waves (NLLS mechanism 3) and the second is the cloud formation upstream of existing clouds (NLLS mechanism 4). For the latter, the existence of clouds lead to a change in the static stability (reduced within the cloud and increased above the cloud) that lifts the NLLJ axis from within the cloud to the height of the cloud top, leading to convergence below the cloud and divergence at cloud top and therefore to further cloud formation via upward transport of cool and moist air. Adler et al. (2017) identified NLLS mechanism 4 as the most important at least for the southern Benin area. However, the comparison with other model studies (e.g. Schuster et al., 2013) is impeded by the different grid mesh size of the realizations.

Adler et al. (2017) identified a regular occurrence of an east-west oriented front followed by a cool airmass that is located stationary 30 km inland along the coast and propagates inland at undisturbed monsoon conditions after 16 UTC. According to the land-sea breeze system, a reversal from sea breeze to land breeze in the evening can be expected. For SWA this reversal is superposed by the stronger southwesterly inland monsoon flow. For example the study of Gbambie and Steyn (2013) indicated for Cotonou (Benin) that the monsoon flow intensifies the sea breeze. Similar characteristics were described in Grams et al. (2010) for Mauritania at the western Atlantic coast. Grams et al. (2010) described this front as a combination of several subsystems that come along with the sea breeze and named it *Atlantic Inflow*. During daytime, the front is stationary at the coast due to the balance between horizontal advection of cool maritime air and the inland turbulence in the PBL that mixes the momentum vertically. The front starts to propagate inland, responding to the SHL pressure-gradient force (Parker et al., 2005a), with the reduction in turbulence in the evening. With the inland propagation, the flow is affected by the Coriolis force and forms a NLLJ (van de Wiel, 2010). In this case the NLLJ can become a continental-scale phenomenon rather than a local feature, as shown by May (1995) for the Australian monsoon.

Grant and van den Heever (2014) focused on the coastal area of Cameroon, denoting the observed front as a sea breeze front. Within an idealized modeling study they assessed the impact of AIE, land-surface characteristics and their synergistic effects on the tropical sea breeze convection. The results indicate a weakening of the sea breeze and less precipitation with increasing aerosol amounts. In the following, we refrain from using the term *sea breeze* for the feature described by Grams et al. (2010) and Adler et al. (2017) since the notation of the land-sea breeze circulation is connected with an onshore wind during daytime and an offshore wind during night. But in case of SWA an onshore wind occurs in the night, too. Instead we introduce the term *Evening Monsoon Flow Enhancement* (EMFE hereafter) which is connected with an EMFE front and an EMFE airmass located behind the EMFE front.

Due to this complex interplay of mechanisms and the problem of identifying the NLLS accurately via remote sensing retrievals (e.g. van der Linden et al., 2015; Hill et al., 2016) also global climate models show serious shortcomings in reproducing the stratus deck over SWA. Hannak et al. (2017) showed that the NLLS is insufficiently represented in global climate models with weak diurnal cycles and a general tendency to underestimate the cloud cover and to overestimate the cloud base height, leading to an overestimation of incoming shortwave radiation at the surface. Based on remote sensing data, Hill et al. (2017) analyzed the effect of different cloud types on the regional energy budget of SWA (cloud radiative effect) for the WAM period June-September of the years 2006-2010. The results indicate that low clouds without second cloud layers in higher altitudes cause a net heating of the surface (as for the other cloud types) but a net cooling of the atmosphere above the low clouds, since in the atmosphere the longwave cooling (-3 W m^{-2}) is stronger than the contribution from shortwave heating (+2 W m^{-2}). By neglecting the low clouds, e.g. if a model cannot reproduce these clouds, Hill et al. (2017) estimated a shortwave cloud radiative effect of -24 W m^{-2} at the top of atmosphere (upward) and 33 W m^{-2} at the surface (downward). The total cloud radiative effect is not dominated by one cloud type but is produced by the sum of several cloud types. Parker et al. (2005a) emphasized the relevance of the coherent diurnal cycle of the monsoon circulation in the lower troposphere for the transport and mixing of trace gases and aerosol. In the daytime convective PBL, air pollutants are vertically mixed while a meridional transport is suppressed. During nighttime, with less vertical mixing and stable stratification, horizontal transport dominates.

Furthermore, the SAL serves as an exchange layer between the monsoon layer and the free troposphere (Parker et al., 2005b). Based on AMMA observations also Matsuki et al. (2010) showed that a strong interaction between the monsoon layer and the SAL exists, leading to entrainment and therefore homogeneous aerosol compositions in both layers, whereas cloud layers disconnect the interaction and lead to different aerosol compositions.

For the period of the DACCIWA measurement campaign in June-July 2016 (Sect. 4.2), Knippertz et al. (2017) provided an overview of the meteorological and air pollution conditions. By focusing on the precipitation difference between the coastal zone (0-7.5°N) and the Soudanian-Sahelian zone (7.5-15°N), they separated the campaign period in four phases that are briefly introduced in the following and subsequently applied in this study. These phases help to assign the DACCIWA-related observations and modeling studies to the general meteorological context and also denote physically meaningful time periods that are appropriate for calculating temporal averages, rather than averaging meteorological or air pollution measures over the entire campaign period. Phase 1 (1 June - 21 June, *Pre-onset phase*) is related to a rainfall maximum near the coast, especially west of Liberia over the Atlantic Ocean, over the Bight of Benin, south of the Niger Delta and over the Bight of Bonny. The ITD is located around 16°N. This phase is related to the lowest spatial coverage of the DACCIWA domain with low-level clouds. Phase 2 (22 June - 20 July, *Post-onset phase*) shows an inland precipitation maximum north of 7.5°N, indicating the establishment of the WAM, linked with a continuously high fraction of low-level clouds (NLLS) over the DACCIWA domain. Knippertz et al. (2017) emphasized that this finding substantiates the hypothesis that the monsoon onset is a preprequisite for the development of NLLS. The ITD fluctuates between 16-22°N. The transition between Phase 1 and 2 is denoted as Phase 1b (16 June - 26 June, *Onset phase*) that is linked with a breakdown of the SHL via the intrusion of cool maritime air from the Atlantic Ocean in the SHL area. Phase 3 (21 July - 26 July, *Wet westerly regime phase*) shows a return of the precipitation maximum to the coast but compared to Phase 1 it is associated with more intense precipitation over the Gulf of Guinea and also precipitation north of 7.5°N in a band between 8°N and 11°N, leading to overall wet conditions. The ITD is located between 20°N and 24°N. Phase 4 (27 July - 31 July, *Recovery phase*) represents the return to the undisturbed monsoon conditions with similar patterns of precipitation but generally higher amplitudes. The coverage with low clouds is as high as in the *Post-onset phase*. The ITD can be found around 20°N. Table 2.2 provides a summary of the meteorological phases describing the DACCIWA campaign conditions. For a more detailed description refer to Knippertz et al. (2017).

Table 2.2: Meteorological phases during the DACCIWA measurement campaign as denoted in Knippertz et al. (2017).

No.	Name	Time Period	Characteristics
1	Pre-onset	1-21 June	Coastal rainfall maximum
1b	Onset	16-26 June	Rainfall maximum shift to continent
2	Post-onset	22 June - 20 July	Undisturbed monsoon conditions
3	Wet westerly regime	21-26 July	Rainfall maximum shift to ocean, overall wet
4	Recovery	27-31 July	Rainfall maximum back to continent

2.2.3 Meteorological Conditions at the Supersites

The meteorological phases denoted in Table 2.2 are reflected in the observations at the SWA supersites during the DACCIWA campaign especially regarding the NLLJ occurrence (Kalthoff et al., 2017). Kalthoff et al. (2017) provided an overview of the observational dataset obtained from the three supersites Savè, Kumasi and Ile-Ife between 14 June and 30 July 2016 during the DACCIWA campaign. Table 2.3 summarizes averaged values of meteorological quantities for Savè and Kumasi. The near-surface temperature shows a range in the diurnal cycle of approximately 6 K, reaching the maximum of 29-30°C around 15 UTC. Precipitation is mostly observed between 15-23 UTC related to convective processes. The monsoon layer has a mean relative humidity (RH) of 80-100% and a mean depth of about 1.9 km AGL. Kalthoff et al. (2017) indicated that the depths of the moist layer and the monsoon layer do not permanently coincide, confirming the above described findings of Parker et al. (2005b) and Matsuki et al. (2010) that an exchange between the monsoon layer and the SAL takes place. The NLLJ develops virtually every night at the supersites with a jet axis around

400 m AGL. Kumasi generally shows a stronger NLLJ and turbulent kinetic energy than Savè, related to topographical features at Savè and a stronger pressure gradient at Kumasi (Kalthoff et al., 2017). The occurrence of NLLS is more distinct in the *Post-onset phase* compared to the *Pre-onset phase* (Fig. 2.5).

Table 2.3: Average meteorological conditions at the supersites Savè (166 m ASL) and Kumasi (266 m ASL), capturing the time period 14 June to 30 July 2016 within the DACCIWA campaign (Kalthoff et al., 2017).

Meteorological quantity	**Savè**	**Kumasi**
Monsoon layer depth (km AGL)	1.9	1.9
Monsoon layer value of wind speed maximum (m s^{-1})	6	8
Monsoon layer height of wind speed maximum (m AGL)	400	400
Monsoon layer RH (%)	80-100	80-100
Nighttime near-surface wind speed (m s^{-1})	1.0	1.5
Nighttime near-surface TKE (m^{-2} s^{-2})	0.2	0.5
NLLS occurrence (%)	64	88
NLLS onset (UTC)	3	0
Cloud base (m AGL)	227	137
Cloud top (m AGL)	587	692
Cloud cover at 12 UTC (%)	80	40
Integrated water vapor (kg m^{-2})	55	50
Total precipitation (mm)	217	258
Sunrise on 3 July (UTC)	05:38	05:56
Sunset on 3 July (UTC)	18:10	18:24
Local time	UTC+1	UTC

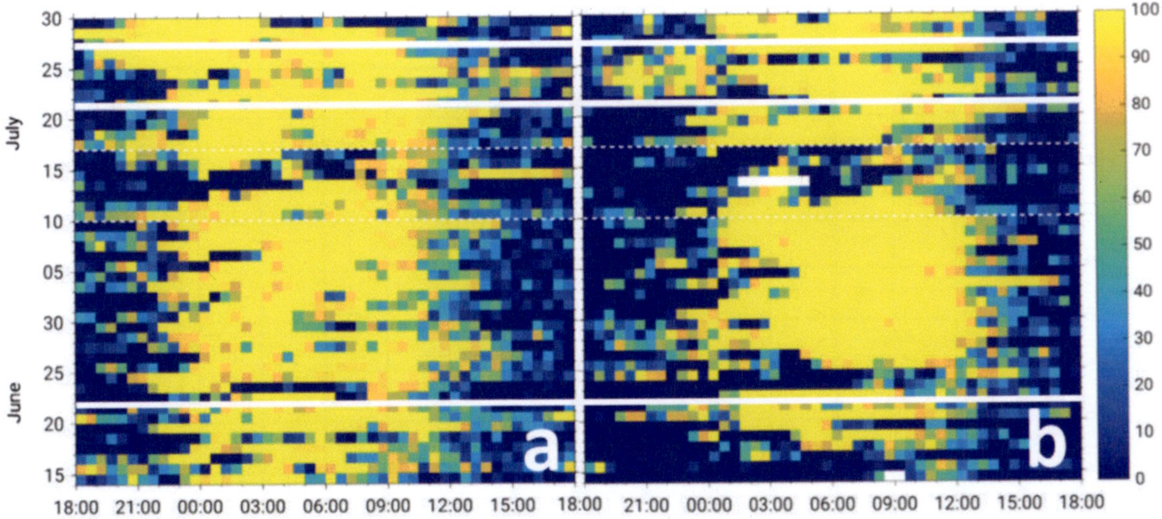

Figure 2.5: Observed 30-min cloud-base fraction (%, 0-1000 m AGL) at the supersites (a) Kumasi and (b) Savè, derived from ceilometer data with a temporal resolution of one minute (Kalthoff et al., 2017, adapted). An observed cloud-base fraction of 100 % is related to 30 consecutive detections of low clouds in the 30 minute interval. The solid white lines separate the meteorological phased defined in Knippertz et al. (2017).

Generally, the cloud diurnal cycle is characterized by a NLLJ maximum between 21-00 UTC, a NLLS onset at 0 UTC (3 UTC) at Kumasi (Savè), a decay of the NLLJ around 6 UTC (sunrise), the beginning of a rise of the cloud base around 7 UTC (one hour after sunrise) and the transition to cumuliform clouds two hours after the beginning of the increase in cloud base height. The NLLS characteristics and temporal evolution show differences within and between the supersites (Kalthoff et al., 2017).

2.3 Atmospheric Composition Characteristics of SWA

The atmospheric composition over SWA is marked by a superposition of local emissions and emissions from remote areas affecting SWA through long-range transports. This includes the emissions of mineral dust, sea salt, biomass burning, biogenic volatile organic compounds (BVOCs) and anthropogenic emissions from cities and industries with the special case of gas flaring emissions from oil industries. Comprehensive insight into the West African atmospheric composition has been achieved within the AMMA project and is summarized in the Atmospheric Chemistry and Physics special issue: AMMA Tropospheric Chemistry and Aerosols. Selected findings of this special issue are summarized in this section.

The Sahara desert is the largest source of mineral dust aerosol in the world with emissions of about 1150 Mt yr^{-1} (Shao et al., 2011). Although the mineral dust tranport to SWA is dominant in boreal winter via the Harmattan, also in the summer monsoon period mineral dust can affect this region. In a study slightly north of the DACCIWA domain, Marticorena et al. (2010) showed that dust events in boreal summer are primarily related to local dust emissions by MCSs. Together with mineral dust, sea salt emissions contribute the largest aerosol mass to the atmosphere (Grythe et al., 2014). During the WAM, sea salt, that is emitted over the Gulf of Guinea, can be transported across SWA within the monsoon layer. Over the eastern equatorial Atlantic Ocean, Grythe et al. (2014) estimated sea spray emissions (primarily consisting of sodium chloride) of 5 kg km^{-2} yr^{-1}. By considering the ocean surface 8°W-8°E and from the equator to the coast, sea spray emissions of $5.2 \cdot 10^{-3}$ Mt yr^{-1} can be expected.

From remote sensing obervations of fire activities it can be seen that biomass burning fires are of less importance in SWA during the WAM, whereas wide areas of central Africa, including Dem. Rep. of the Congo, Angola and Zambia, are characterized by wild fires. Mari et al. (2008) showed that the mid-tropospheric westward transport of biomass burning pollutants from central Africa over the eastern equatorial Atlantic Ocean is controlled by the southern hemispheric easterly jet. With the northward monsoon flow, the pollutants can be transported over SWA limited by the location of the northern hemispheric AEJ. Real et al. (2010) identified biomass burning in central Africa as the source of ozone plumes in the mid and upper troposphere over the Gulf of Guinea.

Vegetated areas are linked with the emission of BVOCs with highest emissions in tropical forest regions (Guenther et al., 2012). As described in Section 2.1, the SWA land surface is covered by evergreen tropical forests in the southern part of Ivory Coast, the southwestern part of Ghana and the southernmost part of Nigeria. Since BVOCs are a precursor for the formation of ozone (Monks et al., 2015) and can form secondary organic aerosol (SOA, Capes et al., 2009), they are able to significantly alter the atmospheric composition of SWA.

Furthermore, the SWA cities depict an important source of pollution from anthropogenic origin. Knippertz et al. (2017) emphasized the variety of these anthropogenic emissions resulting i.a. from transportation, wood and waste burning or charcoal production. Hopkins et al. (2009) estimated the CO, NO_x and volatile organic compound (VOC) emissions of the SWA megacity Lagos (compare Tab. 2.1) from aircraft measurements with 1.44, 0.03 and 0.37 Mt yr^{-1}, respectively. Bahino et al. (2017) measured the trace gases nitrogen dioxide (NO_2), ozone (O_3), ammonia (NH_3), nitric acid (HNO_3)

and SO_2 in Abidjan (Ivory Coast) in winter 2015/2016 (Fig. 2.6). The study reveals a high spatial variability, dominated by the pollution sources near the measurement sites. NH_3 (NO_2) concentrations around 23 (7) ppb were measured in suburban areas, respectively. At domestic fire sites, NH_3 is significantly increased (90 ppb). Highest NO_2 concentrations are found in industrial and traffic areas (20 ppb). The SO_2 concentrations are highest in traffic and suburban areas (1-6 ppb). Bahino et al. (2017) highlight the relevance of the emissions from domestic fires and traffic and the need for an air quality monitoring network in Abidjan.

An additional source of anthropogenic emission is related to the burning of waste gases in the oil producing industry (gas flaring) along the Guina Coast. Nigeria, as the second largest gas flaring country (15 billion cubic meter (bcm) flared gas in 2012) after Russia (Elvidge et al., 2016), is especially affected by these emissions. Deetz and Vogel (2017) developed a method to estimate the spatiotemporal distribution of gaseous compounds from SWA gas flaring. For the Nigerian gas flares, carbon dioxide (CO_2) emissions of 2.7 and 2 Mt yr^{-1} were estimated for 2014 and 2015, respectively. The concept of Deetz and Vogel (2017) is introduced in more detail in Section 3.1.3.

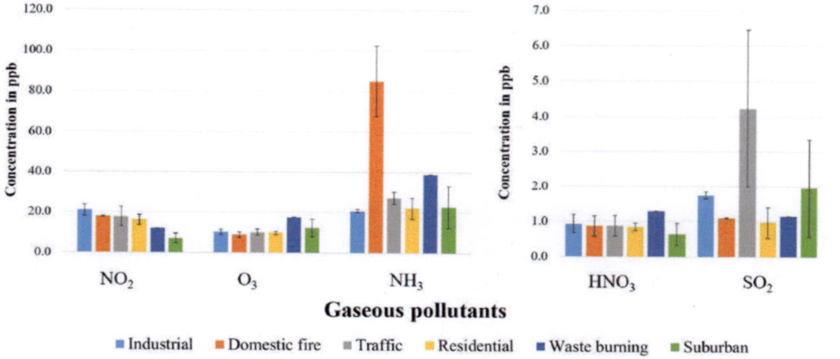

Figure 2.6: Concentration of gaseous pollutants (ppb) separated in anthropogenic emission sectors based on the measurements of 21 sites in the district of Abidjan. The vertical black bars denote the measurement uncertainty (Bahino et al., 2017).

Furthermore, it has to be considered that the African emission from anthropogenic origin will very likely rise in the future (Liousse et al., 2014) leading to even higher burden of air pollutants over SWA.
This set of pollutants from different sources can interact with each other, can be altered by the WAM or could even perturb the WAM itself.
Focusing on the DACCIWA measurement campaign period, Knippertz et al. (2017) assign the atmospheric composition regarding mineral dust, biomass burning CO and urban pollution to the meteorological phases denoted in Table 2.2. They identified strongest mineral dust influence over the DACCIWA domain in the *Pre-onset phase*. The weak southwesterly flow allows a southward propagation of the Saharan dust reaching 6°N. With the beginning of the *Post-onset phase*, the southward mineral dust transport is impeded and restricted to 8°N by the enhanced monsoon flow. With the backward shift to *Pre-onset phase* conditions in the *Wet westerly regime*, that is linked with a cyclone (an anticyclone) north (south) of the DACCIWA domain, an intensified northerly flow is established. This flow transports the mineral dust further northward away from the DACCIWA domain and simultaneously realizes a northward transport of southern hemispheric CO into the DACCIWA domain. In the *Recovery phase*, mineral dust reaches again the northern domain border. Regarding CO, a transport from the southern hemispheric biomass burning areas is visible affecting the entire DACCIWA domain with highest concentrations along the coast.
The city pollution plumes are predominantly transported in northeast direction with the monsoon flow. Since they reach transport distances around 300 km within one day, also inland cities can be affected by emissions from the coastal areas. How the emissions of atmospheric pollutants are considered in this study and implemented in COSMO-ART is described in Section 3.1.3.

2.4 Aerosol-Atmosphere Interactions

In atmospheric science, the basic feedback mechanisms between aerosol and the atmospheric state are well understood, whereas the spatiotemporal interactions between the mechanisms, the interactions between microphysical and dynamical processes and the significance of the interaction with respect to weather and climate are linked to a low level of scientific understanding (Fan et al., 2016; Myhre et al., 2013). For example the modeling study of Ekman (2014) has shown that global near-surface temperature trends between 1965 and 2004 are reproduced most accurately when considering the aerosol amount and supersaturation calculating the cloud droplet number concentration.
The aerosol feedbacks are subdivided in Aerosol Direct Effect (ADE) and Aerosol Indirect Effect (AIE). However, it has to be considered that ADE and AIE are not clearly separated in the atmosphere, leading to synergistic effects or buffering mechanisms (Stevens and Feingold, 2009; Fan et al., 2016).

2.4.1 Aerosol Direct Effect (ADE)

The ADE decribes the interactions between aerosols and the radiation that lead to a modification in the radiative flux due to scattering and absorption by aerosols. This mechanism is approximated via the well-defined Mie theory (e.g. Seinfeld and Pandis, 2006a) and therefore the level of scientific confidence is high (Myhre et al., 2013). However, the assumption of spherical particles is questionable since considerable differences in the optical properties can occur due to the nonsphericity of the particles (e.g. von Hoyningen-Huene and Posse, 1997). At the surface, ADE generally leads to a reduction in shortwave radiation during daytime. During nighttime, a positive longwave ADE can occur due to absorption and the re-emission of longwave radiation from the aerosol layer (e.g. Bangert et al., 2012). The aerosol absorption of shortwave radiation leads to a differential heating of the atmosphere around the aerosol layer, especially in case of black carbon as the most efficiently absorbing aerosol (Peng et al., 2016), affecting the atmospheric stability. This mechanism is referred to as semi-direct aerosol effect that is related to ADE in this study. The current Intergovernmental Panel in Climate Change Assessment Report 5 (IPCC AR5, IPCC, 2013) estimates the global effective radiative forcing by ADE as –0.45 (–0.95 to +0.05) W m^{-2} (Myhre et al., 2013).

Baró et al. (2017) analyzed whether the consideration of ADE in an ensemble of online coupled models improves the 2 m temperature minimum, average and maximum over Europe. The study focuses on a forest fire period and a Saharan mineral dust transport affecting Europe in 2010. The cold bias of the models is reduced by considering ADE or the combination of ADE and AIE in terms of the mineral dust and in the forest fire episode. Generally, the consideration of ADE shows the largest improvements in the representation of the temporal evolution of the temperature with better results for the ensemble than for the single models.

Tao et al. (1996) focused on three longwave radiative mechanisms via a modeling study of a tropical and a midlatitude MCS: (1) The cloud-top cooling and cloud-base warming that might impact the stratification of cloud layers, (2) differential cooling between cloudy and cloud-free areas that might enhance convergence into the cloud system and (3) large-scale radiative cooling that might change the environment. They identified large-scale cooling, linked with an increase in RH, as the dominant process for enhancing the precipitation. Due to the high moisture in the tropics this mechanism is more effective than in midlatitudes. The large-scale cooling increased the tropical rainfall by 36 %. Neglecting the cloud-base warming/cloud-top cooling leads to a reduction in precipitation. The differential cooling shows no clear signal but is identified as less important than the large-scale radiative cooling. The implications of the shortwave radiation reduction to the underlying air-sea coupling is quantified in Chen and Wang (2016). The results of the modeling study with a grid mesh size of 17 km indicated that ADE on the one hand leads to a large (small) decrease in the latent heat flux (sensible heat flux) due to surface cooling and on the other hand to an increase in cloud cover due to the increase in RH in the daytime PBL.

Regarding the AEJ, Tompkins et al. (2005) showed that the use of an improved aerosol climatology for the representation of ADE in the ECMWF leads to an improved representation of the jet strength and location.

The modeling study of Lau et al. (2017) focused on the impacts of aerosol-monsoon interactions on the monsoon variability over the Northern Indian Himalaya Foothills during the summer of 2008. Lau et al. (2017) highlighted that the ADE, i.a. the mineral dust transport and the radiative heating-induced dynamical feedback processes, have major impacts on the large-scale monsoon circulation. The ADE leads to an increased northsouth temperature gradient, a northward displacement of the heavy monsoon precipitation and an advanced monsoon onset over the Himalaya Foothills. The mineral dust leads to an increase in atmospheric stability via the semi-direct effect. The AIE further enhance the ADE by invigorating the deep convection (see Sect. 2.4.2). Lau et al. (2017) underlined the need to consider aerosol-monsoon interactions even in short-term numerical forecasting of the monsoon circulation and precipitation.

2.4.2 Aerosol Indirect Effect (AIE)

The AIE refers to aerosol-cloud interactions. In AR5 the effective radiative forcing from AIE is estimated as –0.45 (–1.2 to 0.0) W m^{-2}. Therefore the total aerosol influence results in a cooling with an effective radiative forcing of approximately -0.9 W m^{-2} (Myhre et al., 2013).

In contrast to ADE, the scientific confidence in AIE, due to their complexity and multitude of counteracting mechamisms, is low (Myhre et al., 2013). The most important AIE mechanism is that aerosol particles can serve as cloud concensation nuclei (CCN) and ice nuclei (IN) leading to the formation of clouds and precipitation (Köhler, 1936; Wegener, 1911). Fan et al. (2016) underlined that the AIE mechanisms furthermore strongly varies between cloud types. In warm shallow clouds the Twomey effect (first AIE) (Twomey, 1977) leads to an increase in cloud albedo with increasing CCN and constant liquid water path due to the reduction in cloud droplet size. The net cloud radiative effect is generally negative on global scale, especially in areas of extensive low-level stratus and stratocumulus (Boucher et al., 2013). For these clouds, the shortwave cloud radiative effect is strong (high albedo) but the longwave cloud radiative effect is weak (warm cloud top).

The Albrecht effect (second AIE) (Albrecht, 1989) describes the increase in cloud lifetime due to the inhibition of precipitation with increasing CCN. Rosenfeld (1999) demonstrated the precipitation inhibition due the Albrecht effect via the remote sensing detection of precipitation and biomass burning aerosol over Indonesia.

For deep convective clouds the mechanisms are less well understood. The convective-cloud invigoration mechanism, proposed by Rosenfeld et al. (2008), states that the cloud water surplus from warm rain suppression is transported to higher levels and leads to an intensification of convection due to latent heat release from freezing. However, other modeling studies reveal opposite signals (e.g. Fan et al., 2013). Koren et al. (2014) proposed that the convective-cloud invigoration mechanism is a special case of the concept of aerosol-limited clouds (cloud development is limited by the availability of CCN). Therefore the invigoration already occurs in pure warm clouds, when the condensation of water on the available CCN leads to latent heat release and a subsequent upward motion. Koren et al. (2014) furthermore suggested that in many cases the convective clouds are even aerosol-limited to higher aerosol amounts up to anthropogenic levels. This is in agreement with the findings of Sheffield et al. (2015) focusing on the AIE on cumulus congestus clouds.

Bangert et al. (2012) assessed the impact of mineral dust on the ADE and AIE by using the regional-scale model COSMO-ART. They modeled a long-range transport of Saharan dust to Europe indicating a reduction of incoming shortwave radiation up to -75 W m^{-2} and surface temperature reductions of -0.2 to -0.4 K. Furthermore, the results underline that the impact on cloud droplet number concentrations is considerably smaller than the impact on ice crystal number concentrations in clouds within a temperature range of 0 to -38°C. Neubauer et al. (2014) analyzed the anthropogenic aerosol effect in marine stratocumulus clouds with the global climate model ECHAM6-HAM2. The study reveals that the anthropogenic aerosol has only a small impact on the stratocumulus cloud properties but reduces the globally averaged effective radiative forcing from -1.08 to -1.19 W m^{-2}.

Among the relevance of the aerosol as CCN, also the role as IN is subject to current research. However, this study focuses on the warm clouds and therefore excludes this field of research. Further information about the IN effects on AIE can be found in Fan et al. (2016).

Due to several reasons, the main focus of AIE research is located on low-level warm clouds. They cover large areas and are linked to less complex microphysics since no mixed-phase or cold phase has to be considered in AIE. In geoengineering research, these clouds are identified as a pathway to reduce global warming by artificially increasing the cloud albedo (Latham et al., 2008). This process is realized unintentially and can be observed along ship tracks. The aerosol increase from shipping emissions leads to cloud brightening, precipitation suppression and an increase in cloud water related to the Twomey- and Albrecht effect (e.g. Lu et al., 2007). According to the above-mentioned mechanisms, the aerosol increase in warm clouds lead to an increase in the cloud albedo, furthermore the cloud droplet size decreases, leading to a reduced efficiency of collision and coalescence and with that to less warm rain and to a longer cloud lifetime. While cumuliform clouds are formed by surface heating, stratus and stratocumulus clouds are formed by cloud-top radiative cooling. Studies of shallow cumuli over land indicate an increase (a decrease) in cloud droplet number (cloud droplet size) with increasing aerosol (e.g. Shrivastava et al., 2013). Xue and Feingold (2006) emphasize buffering mechanisms that have to be considered, e.g. that the increase in aerosol number on the one hand increase the cloud water by suppressing rain but on the other hand increase evaporation and entrainment. Saleeby et al. (2014) extended this buffering mechanism to the vertical motion with increased upward motion due to latent heat release from condensation and increased downward motion due to the cooling from evaporation and entrainment.

A typical feature of warm clouds is the transition from stratus clouds in the early morning (radiative cooling) to late-morning convective clouds (thermal heating). Kruger et al. (2015) proposed four stages in the stratus-to-cumulus transition: the stratus-topped boundary layer, the well mixed deep stratus-topped boundary layer, the cumulus under stratocumulus boundary layer (double layer structure) and the trade cumulus boundary layer.

Saleeby et al. (2014) studied the AIE in several marine warm-cloud regimes. In addition to the above-mentioned mechanisms, they highlight the increased evaporation and entrainment at the inversion layer (stratus top) with higher aerosol amounts (smaller cloud droplets), leading to a reduction in stratus cloud water. Deeper penetrating cumuli within the stratus and with a cloud top above the stratus are thinning the surrounding stratus layer by entrainment and subsequent subsidence of colder air. From this, it can be concluded that the increase in CCN promotes and accelerates the stratus-to-cumulus transition. Furthermore, the cloud base increases since the increasing aerosol reduces the low-level cumuli below the stratus (less cumuli in the cumulus under stratocumulus boundary layer). Saleeby et al. (2014) related this cumuli reduction to the following reaction chain: higher aerosol amount leads to less precipitation, less latent cooling, less downdrafts, less surface convergence and therefore to less updrafts and development of cumuli. They identified cloud patterns with an increased occurrence of clouds with low and high liquid water path and a decrease of clouds with moderate liquid water path with increasing aerosol, leading to a 50 % reduction in the total domain rainfall but via the convective-cloud invigoration mechanism to greater local rainfall amounts. This study emphasizes that the AIE mechanisms includes the linkage between microphysical and dynamical components but also between several coexisting cloud types. The observational study of McComiskey et al. (2009) assessed the AIE on stratiform clouds at the Californian coast, revealing that changes in the aerosol amount accounts for 14 % of the variation in the cloud microphysical properties. They furthermore emphasize that the AIE has to be discussed in relation to the cloud water path. For low cloud water values, AIE reflects the combination of activation and collision/coalescence and for high cloud water amounts (above 150 g m^{-2}) the strong variability in AIE is rather related to precipitation scavenging of aerosol and therefore the impact of clouds on aerosol.

In a study of the Indian monsoon, Lau and Kim (2006) indicated that the high absorbing aerosol amount over the Indian subcontinent, that is linked with a heating of the upper troposphere, leads to an advance of the onset and an increase in the monsoon precipitation in that area. Koren et al. (2012) extended the study of Rosenfeld (1999) to a global scale to identify impacts from AIE by using remote sensing data of precipitation and AOD as a proxy for aerosol for June-August 2007. Over large areas with significantly different cloud systems, an increase in AOD lead to an increase in precipitation and cloud top height. For the WAM region, a coherent pattern of decrease in precipitation (-0.3 mm h^{-1}) with increasing AOD over the Gulf of Guinea and an increase in precipitation over the SWA inland area up to 10°N (+0.3 mm h^{-1}) can be observed.

The interaction between AIE, the land surface characteristics and tropical sea breeze convection over Cameroon was analyzed in Grant and van den Heever (2014) for boreal summer conditions. The study reveals a weakening of the sea breeze front with increasing aerosol, linked with less precipitation, due to a reduction in surface shortwave radiation and therefore surface heating. A reduction in soil moisture leads, via the reduction in the near-surface latent heat flux, to less available moisture for convection and therefore also to less precipitation. Grant and van den Heever (2014) highlight the importance of synergistic effects between the aerosol amount and land surface characteristics. Fan et al. (2016) identified the key issues that impede an improved understanding of AIE in terms of modeling i.a. in the insufficient model representation of the updraft intensity and cloud properties, the difficulties in differentiating between the actual aerosol impacts and the natural variability, the cloud microphysical parameterization and the subgrid-scale cloud parameterization in Global Circulation Models. Stevens and Feingold (2009) and Fan et al. (2016) emphasized the need to analyze AIE cloud-regime dependent with fine-scale models to explicitly resolve the interacting processes rather than using global models with parameterizations. These suggestions become even more relevant when focusing on the complex characteristics of the WAM. For example Marsham et al. (2013) indicated that the WAM representation shows fundamental differences in the UK Met Office Unified Model between the consideration of explicit and parameterized moist convection.

From the literature review in terms of the aerosol-atmosphere interaction, four aspects become apparent: (1) The aerosol impact on the WAM is less represented in the existing studies than e.g. for the Asian Monsoon, (2) there are more aerosol studies focus on AIE than on ADE although ADE very likely have stronger impacts on the atmospheric dynamics, (3) the studies of AIE and ADE, when focusing on warm cumuliform and stratiform clouds, primarily concentrate on marine clouds and (4) it poses a challenge to disentangle AIE from the system of coupled feedback mechanisms.

The present study refers to these four aspects and intends to provide added value by (1) focusing on the WAM, (2) emphasizing the relevance of ADE on the atmospheric dynamics, (3) explicitly focusing on warm clouds over land and (4) analyzing AIE with COSMO-ART with the support of the comprehensive DACCIWA observations. We are not aware of previous studies, focusing with a regional-scale model on SWA by considering ADE and AIE. Furthermore, COSMO-ART was applied for the first time to the tropics.

Referring to Stevens and Feingold (2009), this study is positioned, by depicting a highly resolved model study of a specific cloud regime, as a contribution to the overall understanding of different cloud regimes.

3 Methods

This section comprises an overview of the methods used in the present study. The key method is the model framework COSMO-ART that is described in Section 3.1. This study has explicitly contributed added value to the scientific work via the further developments of the atmospheric pollutant emission parameterizations in COSMO-ART that are highlighted in Section 3.1.3.

3.1 The Numerical Model Framework of COSMO-ART

For this study, the regional-scale model framework COSMO-ART (Consortium for Small-scale Modeling - Aerosols and Reactive Trace gases, Vogel et al., 2009) is used. COSMO-ART is based on the operational weather forecast model COSMO of the German Weather Service (DWD, Baldauf et al., 2011), that is introduced in Section 3.1.1. The ART extensions, described in Section 3.1.2, allow for an online treatment of the aerosol dynamics and atmospheric chemistry. Knote et al. (2011) validated the aerosol and gaseous compounds in detail against observations for the European area. For this study, new comprehensive parameterizations for the representation of natural and anthropogenic emissions of aerosols and trace gases were implemented to further develop the model and to apply COSMO-ART to the conditions of SWA. Section 3.1.3 refers to the emission parameterizations highlighting the new developments. Furthermore, COSMO-ART allows to consider the ADE

and AIE that are essential to achieve the objectives of this study.
The full capacity of COSMO-ART was applied in numerous studies. Stanelle et al. (2010) analyzed the ADE of mineral dust over the Sahara that alters the near-surface temperature up to 4 K in case of elevated mineral dust layers. Furthermore, feedbacks between the mineral dust ADE and the atmospheric dynamics lead to modifications in the mineral dust emission. Athanasopoulou et al. (2014) quantified a severe wildfire event over Greece in 2007 in terms of air quality and ADE that reveals AOD values between 0.75 and 1 and a cooling of 0.5 K. Walter et al. (2016) extended COSMO-ART with a plume-rise model to describe biomass burning pollution injection heights and applied the model to Canadian forest fires in 2010. The ADE related to the biomass burning plume leads to a near-surface cooling of up to 6 K. The ADE of sea salt over the Mediterranean Sea, Northeast Atlantic, North Sea and Baltic Sea was modeled by Lundgren et al. (2013) in accordance to the observations from remote sensing. Kraut (2015) applied an ensemble approach by including random temperature pertubations to isolate the sea salt AIE on the characteristics of a cyclone over the Mediterranean Sea in 2011, revealing spatial shifts and intensity differences in the precipitation patterns. By considering the AIE on post-frontal convective clouds over Germany in 2008, the cloud properties were changed significantly, leading to a reduction in precipitation with increasing aerosol amounts (Rieger et al., 2014).
The model representation of the feedback mechanisms between aerosol, radiation and clouds is described in Section 3.1.4.

3.1.1 The Regional Atmospheric Model COSMO

COSMO is a nonhydrostatic regional-scale atmospheric prediction model used by the DWD for numerical weather prediction over Germany with a grid mesh size of 2.8 km (COSMO-DE). COSMO-EU covering Europe was replaced by ICON-EU on 30 November 2016 using the ICON model of the DWD

(Zängl et al., 2015). COSMO is based on the primitive thermo-hydrodynamical equations describing the compressible flow in a moist atmosphere (Doms and Baldauf, 2015). The equations are using rotated geographical coordinates (Arakawa C-grid) and a generalized terrain-following height coordinate.
The cloud microphysics in COSMO is described via a bulk water-continuity model (one-moment microphysics scheme, OMMS) (Doms et al., 2011). In this case, the mass density (3rd moment of the size distribution) of the hydrometeors water vapor q_v, cloud water q_c, cloud ice q_i, rain q_r, snow q_s and graupel q_g are predicted (referring to the *Graupel Scheme* of COSMO). The use of moments to characterize a size distribution is introduced in more detail in Section 3.1.2. These hydrometeors are further separated in precipitating (q_r, q_s, q_g) and non-precipitating particles (q_c, q_i). For these two categories the budget equations were calculated. Regarding the condensation rate, a saturation equilibrium with respect to water in clouds is assumed. This is realized by a saturation adjustment in which the concentrations of q_v and q_c are instantaneous adjusted to a saturated state in case a grid box reaches supersaturation. This adjustment changes the specific humidity that is used as the amount of water which is condensed (in case of supersaturation) or evaporated (in case of subsaturation), by assuming that enough CCN are available at all times.
Seifert and Beheng (2006) developed a two-moment microphysics scheme (TMMS). In addition to the mass density of the hydrometeors (3rd moment of the size distribution) also the particle density (0th moment of the size distribution) is predicted. It considers nucleation, condensational growth, the formation of raindrops from cloud droplet coagulation (autoconversion), coagulation without switching the drop category (selfcollection), collisional breakup of raindrops, collection of cloud droplets through raindrops (accretion) and evaporation as well as sedimentation of raindrops regarding the warm phase and nucleation, depositional growth, freezing, melting, riming and aggregation for the cold phase (Seifert and Beheng, 2006). For the nucleation in the warm and cold phase the aerosol characteristics are crutial. Without considering the prognostic aerosol of COSMO-ART, TMMS uses aerosol

climatologies for maritime and continental conditions. In case of allowing the interaction between the aerosol distributions simulated by COSMO-ART and the cloud microphysics in TMMS, additional parameterizations of the ART model part were used, which are described in Section 3.1.4. In this case, the TMMS represents the interface between the atmospheric state in COSMO and the online-calculated aerosol in COSMO-ART to describe the AIE.

3.1.2 Treatment of Aerosols and Trace Gases in COSMO-ART

Modal Approach

The number density of an aerosol population can be described by a continuous distribution function $n(d_p)\, dd_p$ with the particle diameter d_p. The integral over all size intervals gives the total number density N

$$N = \int_0^\infty n(d_p)\, dd_p;\ [N] = \frac{Number}{cm^3} \tag{3.1}$$

Analog a formulation for the total surface area density S, volume density V and mass density M is possible when assuming spherical particles:

$$S = \int_0^\infty \pi d_p^2\, n(d_p)\, dd_p;\ [S] = \frac{\mu m^2}{cm^3}, \tag{3.2}$$

$$V = \int_0^\infty \frac{1}{6}\pi d_p^3\, n(d_p)\, dd_p;\ [V] = \frac{\mu m^3}{cm^3} \text{ and} \tag{3.3}$$

$$M = \int_0^\infty \frac{1}{6}\pi \rho_p d_p^3\, n(d_p)\, dd_p;\ [M] = \frac{\mu g}{cm^3}, \tag{3.4}$$

with the particle density ρ_p.
Via M_k

$$M_k = \int_0^\infty d_p^k\, n(d_p)\, dd_p \tag{3.5}$$

the k-th moment of the distribution $n(d_p)\, dd_p$ can be defined. From Equation 3.5 the total number density is proportional to M_0 with $k = 0$ (0th moment), the surface area density is proportional to M_2 with $k = 2$ (2nd moment) and the volume and mass density is proportional to M_3 with $k = 3$ (3rd moment)

$$N = M_0, \tag{3.6}$$

$$S = \pi\, M_2, \tag{3.7}$$

$$V = \frac{1}{6}\pi\, M_3 \;\; and \tag{3.8}$$

$$M = \frac{1}{6}\pi \rho_p\, M_3. \tag{3.9}$$

In COSMO-ART the aerosol distributions are represented as a sum of lognormal distributions (modes herafter), following Whitby (1978), that are defined as

$$n_l(ln\, d_p) = \frac{N_l}{\sqrt{2\pi}\, d_p\, ln\sigma_{g,l}}\, exp\left(-\frac{(ln\, d_p - ln\, d_{pg,l})^2}{2\, ln^2\sigma_{g,l}}\right), \tag{3.10}$$

where l specifies the mode and by using the logarithm of the diameter within the distribution function. Lognormal distributions (Aitchison and Brown, 1957) on the one hand match well the observed shapes of aerosol size distributions and are therefore a good approximation and on the other hand are uniquely described by only three parameters, the total number density of mode l, N_l, the median diameter $d_{pg,l}$ and the geometric standard deviation $\sigma_{g,l}$, which simplifies the treatment of these distributions in atmospheric models. The expression in the exponent of Eq. 3.10 denotes a numerical value equation and therefore only the numerical value of d_p and $d_{pg,l}$ is substituted but not the unit. The total aerosol distribution given as

$$n(ln\, d_p) = \sum_{l=1}^{L} \frac{N_l}{\sqrt{2\pi}\, d_p\, ln\sigma_{g,l}} exp\left(-\frac{(ln\, d_p - ln\, d_{pg,l})^2}{2\, ln^2\sigma_{g,l}}\right), \tag{3.11}$$

where L is the number of modes. As for Eq. 3.10 the expression in the exponent of Eq. 3.10 denotes a numerical value equation and therefore only the numerical value of d_p and $d_{pg,l}$ is substituted but not the unit. For COSMO-ART three modes are considered: Aitken, accumulation and coarse mode ($L = 3$). According to Equation 3.1 and by using 3.10, the total number density is the sum over the single modes

$$N = \int_0^\infty n(ln\, d_p)\, d(ln\, d_p). \tag{3.12}$$

The temporal evolution of the aerosol size distribution is in the following not calculated by the prognostic equations of the three parameters of the mode. Whitby et al. (1991) showed that this method is not possible when considering several processes (e.g. coagulation and diffusion) simultaneously. Alternatively, three moments are introduced via the Equations 3.6-3.9. Substituting Equation 3.10 into 3.5 delivers

$$M_k = \int_0^\infty d_p^k\, n(ln\, d_p)\, d(ln\, d_p) = N\, d_{pg}^k\, exp\left(\frac{k^2}{2}\, ln^2\, \sigma_g\right). \tag{3.13}$$

The aerosol dynamic processes in COSMO-ART are treated in the aerosol module MADEsoot (Modal Aerosol Dynamics Model for Europe extended by soot, Vogel et al., 2009), which is based on the aerosol module MADE of Ackermann et al. (1998). Within COSMO-ART, 12 modes are defined, considering SOA and the directly emitted aerosols soot, mineral dust and sea salt. Table 3.1 summarizes the modes with their chemical composition, initial mean diameter and standard deviation.

Table 3.1 *(facing page)*: Chemical composition, initial mean diameter and standard deviation of the 12 aerosol modes in COSMO-ART (AIT: Aitken mode, AITS: Aitken mode containing a soot core, ACC: accumulation mode, ACCS: accumulation mode containing a soot core, SOOT: pure (fresh) soot mode, CANTHRO: coarse mode of anthropogenic origin, CSEAS: coarse mode of marine origin (sea salt fine (CSEASA), sea salt medium (CSEASB) and sea salt coarse (CSEASC)) and CSOIL: coarse mode of mineral origin (mineral dust fine (CSOILA), mineral dust medium (CSOILB) and mineral dust coarse (CSOILC)).

	AIT	AITS	ACC	ACCS	SOOT	CANTRHO	CSEASA	CSEASB	CSEASC	CSOILA	CSOILB	CSOILC
Soot		•		•	•							
Sulfate (SO_4^{2-})	•	•	•	•			•	•	•			
Ammonium (NH_4^+)	•	•	•	•								
Nitrate (NO_3^-)	•	•	•	•								
Organics	•	•	•	•								
Water	•	•	•	•			•	•	•			
Sodium chloride							•	•	•			
Initial diameter (μm)												
Number	0.01	0.08	0.07	0.08	0.08	1	0.2	2	12	0.64	3.5	8.7
Mass	0.03	0.03	0.3	0.3	0.17	6	0.69	0.85	27.93	1.5	6.7	14.2
Standard deviation												
	1.7	1.7	2.0	2.0	1.4	2.5	1.9	2.0	1.7	1.7	1.6	1.5

The latter is kept constant while the mean diameter changes within the transport processes. Sea salt (CSEAS) and mineral dust (CSOIL) are described by a sum of three lognormal modes, respectively. They represent the coarse mode of marine and mineral origin, respectively. This is completed by the coarse mode of anthropogenic origin PM10 (CANTRHO).

The submicron particles are described via five lognormal modes, in which AIT and AITS represent the Aitken mode for secondary particles consisting sulfate, ammonium, nitrate, organics and water (AIT) and aged soot particles consisting of sulfate, ammonium, nitrate, organics and water in an internal mixture (AITS), which is analog for the accumulation mode (ACC and ACCS). SOOT represents the pure (fresh) soot mode. The modes SOOT, AITS and ACCS were added to MADE by Riemer (2002) to represent the aging of soot particles. Exchange between the submicron modes are possible through condensation and coagulation. Between the submicron modes and the coarse modes no interactions are intended in COSMO-ART. Therefore these modes are decoupled.

With the fixed geometric standard deviation (see Tab. 3.1) only two moments per mode have to be derived prognostically to finally derive the temporal evolution of the aerosol number size distribution. In COSMO-ART these are the 0th moment (number density, $k=0$) and 3rd moment (mass density, $k=3$). The prognostic equation of the moments is derived by using the prognostic equation of the integral moments M_k of Friedlander (1977) and Equation 3.11 by applying a Reynolds averaging. The moments M_0 and M_3 are normalized with the total number density of air molecules and the total mass density of humid air, respectively (Vogel et al., 2009). Exemplarily, the prognostic equation for M_0 regarding AIT is presented:

$$\frac{\partial \hat{M}_{0,AIT}}{\partial t} = \underbrace{-\hat{\mathbf{v}} \cdot \nabla \hat{M}_{0,AIT}}_{1} \underbrace{-\overline{v}_{sed,0,AIT} \frac{\partial \hat{M}_{0,AIT}}{\partial z}}_{2}$$

$$\underbrace{+\frac{1}{\bar{\rho}_a}\nabla\cdot F^{M_{0,AIT}}}_{3}\underbrace{-\overline{W}_{0,AIT}}_{4}\underbrace{-\overline{Ca}_{0,AIT-AIT}}_{5}\underbrace{-\overline{Ca}_{0,AIT-ACC}}_{6}$$

$$\underbrace{-\overline{Ca}_{0,AIT-AITS}}_{7}\underbrace{-\overline{Ca}_{0,AIT-ACCS}}_{8}\underbrace{-\overline{Ca}_{0,AIT-SOOT}}_{9}\underbrace{+\overline{Nu}_{0}}_{10}. \quad (3.14)$$

Equation 3.14 emcompass the terms (1) advection with the mean wind speed $\hat{\mathbf{v}}$, (2) sedimentation with the mean sedimentation velocity of the 0th moment regarding AIT, $\overline{v}_{sed,0,AIT}$, (3) turbulent diffusion with the turbulent fluxes of the 0th moment regarding AIT, $F^{M_{0,AIT}} = \overline{\rho\, v''\, M''_{0,AIT}}$, (4) wet deposition, (5) intramodal coagulation within AIT, (6-9) intermodal coagulation between AIT and ACC, AITS, ACCS and SOOT, respectively and (10) nucleation. From the different signs it can be seen that (2) and (4-9) are sinks that result in a reduction in the number density and (10) is a source of aerosol particles. The processes (1) and (3) can both decrease or increase the number density. For each mode the prognostic equations for the number and mass density are solved. The substances sulfate (SO_4^{2-}), nitrate (NO_3^-), ammonium (NH_4^+) and water in the submicron modes are in thermodynamic equilibrium. Therefore only the prognostic equation for the mass density of SO_4^{2-} is calculated and the aerosol distributions of NH_4^+ and NO_3^- are derived via the equilibrium theory (Fountoukis and Nenes, 2007). The full set of prognostic equations is given in Vogel et al. (2009).

The aerosol distribution, as shown in Equation 3.14, can be altered via advection, nucleation, coagulation, turbulent diffusion, sedimentation and dry deposition, wet deposition, condensation, aging and emission. Condensation only changes the mass density but not the number density. Direct emissions are only considered for SOOT and the marine, mineral and anthropogenic contributions of the coarse mode. Within a mode, the particles are internally mixed (uniform particle composition) (Riemer, 2002). In the following, the representation of the aerosol dynamics mechanisms in COSMO-ART are briefly introduced.

Aerosol Chemistry

COSMO-ART treats the aerosol chemistry for sea salt via Lundgren (2010) and for anthropogenic aerosol via Fountoukis and Nenes (2007). Mineral dust is treated as an inert aerosol in COSMO-ART and is therefore not subject to aerosol chemistry. However, mineral dust can act as CCN and IN and is subject to sedimentation. Via Lundgren (2010) the sea salt aerosol chemical composition in COSMO-ART can be altered by condensation of sulfuric acid (H_2SO_4) onto sea salt aerosol and by the uptake of water. Sources of sulfur are oceanic dimethyl sulfide (DMS) (see Section 3.1.3) and anthropogenic emissions from ships or land areas.
The aerosol liquid water content (ALWC, kg water m^{-3} air) depends on RH and the chemical composition of the aerosol (molarity). For a RH near to 100% the ALWC strongly increases whereas a decrease in RH leads to evaporation of the ALWC. Below an aerosol specific RH the aerosol crystallizes and water uptake stops. The ALWC is determined via the so called ZSR relation (Zdanovskii, 1948; Stokes and Robinson, 1966)

$$ALWC = \sum_i \frac{M_i}{m_{0,i}(a_w)}, \tag{3.15}$$

where M_i is the number of moles of the species i per air volume and $m_{0,i}$ the number of moles of species i per kg solvent (water) with the water activity $a_w = \frac{RH}{100}$. The ALWC calculation in COSMO-ART assumes that the water uptake of the particles does not alter the ambient relative humidity, neglects the Kelvin effect and furthermore water uptake only takes place in sulfate-poor environments since in sulfate-rich environments the water is completely consumed by the formation of H_2SO_4. In COSMO-ART, the sea salt ALWC is zero below 47% RH, increases with a RH above 47% and remains constant for a RH greater than 99% (Lundgren, 2010).
For the consideration of aerosol chemistry regarding the anthropogenic aerosol, the thermodynamic equilibrium model ISORROPIA II (Fountoukis and Nenes,

2007) is used in COSMO-ART. ISORROPIA II treats the chemical equilibrium between the gas, liquid and solid phase for the aerosol system containing the inorganic ions potassium (K^+), calcium (Ca^{2+}), magnesium (Mg^{2+}), NH_4^+, natrium (Na^+), SO_4^{2-}, NO_3^-, chloride (Cl^-) and water (H_2O).
H_2SO_4 has a very low vapor pressure and therefore is preferentially in solid phase, either through condensation on preexisting particles or by the formation of new particles via nucleation. In contrast, NO_3^- and NH_4^+ have significant higher vapor pressures and therefore can be found both, in gas and solid phase. The distribution depends on the concentration of SO_4^{2-}, NO_3^-, RH and temperature. ISORROPIA II consists of 27 equilibrium equations to derive equilibrium concentrations of the species. The resulting concentrations of SO_4^{2-}, NH_4^+, NO_3^- and H_2O were distributed to the corresponding modes according to Table 3.1. As for sea salt, the ALWC for anthropogenic aerosol is derived via Equation 3.15. However, the determination is more complex due to the multicomponent mixture. A detailed description of the aerosol chemistry in ISORROPIA II can be found in Fountoukis and Nenes (2007).
The ALWC can largely contribute the the total aerosol volume in regimes with RH above 90% (Bian et al., 2014), modifying the aerosol optical properties, the AOD and therefore the strength of the ADE. This is especially relevant for this study with a research focus on the WAM with its warm and moist monsoon layer.

Nucleation

The formation of SOA from the gas phase is calculated via a volatility basic set (VBS) approach (Athanasopoulou et al., 2013). The precursors (VOCs) are higher alkanes, terminal and internal alkenes, toluene, xylene and cresol for anthropognic and isoprene, α-pinene and limonene for biogenic. Within COSMO-ART these VOCs undergo oxidation reactions which deliver semi-volatile components, grouped according to their effective saturation

concentration C* (at 298 K) 1 μg m^{-3} (SOA1), 10 μg m^{-3} (SOA10), 100 μg m^{-3} (SOA100) and 1000 μg m^{-3} (SOA1000). Even after the transformation of the precursors to SOA, further SOA formation is possible via reactions with hydroxyl (OH^-) radicals.
Secondary inorganic particles are formed in the modes AIT and ACC via the nucleation H_2SO_4 and water (binary homogeneous nucleation). The nucleation rates are calculated with the parameterization of Kerminen and Wexler (1994). The primary source of H_2SO_4 in the atmosphere is the oxidation of SO_2 via OH^- radicals. A partitioning of the inorganic aerosol between the gas phase and the particulate phase is modeled via ISORROPIA II (Fountoukis and Nenes, 2007).
Soot particles of mode SOOT, primary organics in mode CANTHRO as well as sea salt and mineral dust originate from direct emissions. Regarding the modes AIT, AITS, ACC and ACCS no direct emissions of sulfate and primary organics are considered.

Condensation

The aerosol distribution changes when a vapor condenses on or evaporates from aerosol particles and alters thereby the particle diameter. All submicron particle modes are altered by condensation, which affects the aerosol mass density but not the number density. As for the inorganic nucleation, also for the condensation only H_2SO_4 is considered in COSMO-ART. The particle growth rate of the mass density due to condensation is derived via Whitby et al. (1991),

Coagulation

Aerosol particles that are suspended in a fluid (e.g. air) can contact other particles due to their Brownian motion (thermal coagulation). Other processes causing coagulation are e.g. the coagulation in laminar shear, turbulent flow or due to gravitational settling. All submicron particle modes in COSMO-ART are subject to coagulation. The particles are assigned to the different modes according to Whitby et al. (1991). While particles that are formed by intramodal coagulation stay in their original modes, the particles formed by intermodal coagulation belonging afterwards to the mode with the larger median diameter. In COSMO-ART, only the coagulation due to Brownian motion is considered. Furthermore, it is assumed that the aerosol distribution stays lognormal despite the coagulation. The intermodal coagulation changes the number density but not the mass density in the corresponding mode.

Soot Aging

As indicated in the introduction of the modal approach of COSMO-ART, Riemer (2002) implemented the three additional submicron modes SOOT, AITS and ACCS to represent the aging of soot particles within MADE. The implementation, including the extended balance equations for the 0th and 3rd moment, is given in Riemer (2002).

Weingartner et al. (1997) showed that freshly emitted soot particles are hydrophobic and therefore less suitable as CCN. During the aging process the hygroscopicity can change (more hydrophilic) including the transfer of soot into an internal mixture as observed by Hughes et al. (2000) for the Los Angeles Basin. The hygroscopicity determines the aerosol growth in environments with high RHs. On the one hand the coating of soot with soluble substances influences the optical properties by increasing the absorption (Fuller et al., 1999). Via the cloud albedo reduction this could lead to a warming. On the other hand the soot particles, that are more hydrophilic after aging, are more suitable as CCN and can increase the cloud albedo via the Twomey effect, which could lead to a cooling. Both effects have implications on the energy balance (Jacobson, 2000; Riemer et al., 2004) and the atmospheric residence time of the soot particles (Dlugi, 1989). The aging of soot particles is related to the following mechanisms: coagulation, which transforms the externally mixed soot to internal mixtures, condensation of hydrophilic organic or inorganic substances on soot surfaces and photochemical reactions at the soot surface (not considered in MADEsoot).

The coagulation of soot particles in SOOT with particles in AIT, AITS, ACC or ACCS transfers the mass of SOOT into the modes AITS or ACCS. Additionally, condensation of inorganic (H_2SO_4 with subsequent formation of ammonium sulfate) and organic (SOA) substances at the soot particle surface can transfer soot into an internal mixture as well. Based on Weingartner et al. (1997) all particles of SOOT are moved to AITS and ACCS if the soluble mass density fraction of SO_4^{2-} in SOOT is greater than 5 %.

Turbulent Diffusion

For describing the vertical turbulent diffusion the 3D TKE-based prognostic scheme was selected in COSMO which is described in detail in Doms et al. (2011).

Advection

For the tracer transport, the mass conserving advection scheme of Bott (1989) was selected within COSMO.

Sedimentation and Dry Deposition

The sedimentation and dry deposition of aerosols is parameterized according to Binkowski and Shankar (1995) and Ackermann et al. (1998). The sedimentation velocity for a polydispersed aerosol is derived by averaging the sedimentation velocity for a monodispersed aerosol over the kth moment of the aerosol size distribution and assuming a lognormal aerosol size distribution. In COSMO-ART the two moments $k = 0$ (number density) and $k = 3$ (volume density regarding the modes AIT, AITS, ACC, ACCS and SOOT and mass density regarding CSEASA, CSEASB, CSEASC, CSOILA, CSOILB and CSOILC) were used. The sedimentation velocity is derived for all model layers and the deposition velocity just for the lowest model layer.

Wet Deposition

The wet deposition is parameterized after Rinke (2008) and depends on the aerosol and precipitation particle size distribution. Wet deposition summarizes the two main mechanisms in-cloud scavenging (rainout), when activated particles grow to cloud droplets, fall to the ground and deposit the aerosol, and below-cloud scavenging (washout). Rinke (2008) focused on the below-cloud scavenging for the implementation in COSMO-ART. In-cloud scavenging is not considered in the model. The latter is related to two processes, the nucleation scavenging (growth of CCN to cloud droplets) and the subsequent collection of interstitial aerosol (aerosol that is to small to activate). In clouds that are influenced by anthropogenic sources, the nucleation scavenging efficiencies are small and only exceed 0.1 in the remote marine atmosphere (Seinfeld and Pandis, 2006b). Therefore the neglection of in-cloud scavenging is justifiable for the application to SWA. The precipitation particle (raindrop) size distribution is described after Mircea and Stefan (1998).

Emission

The description and parameterization of the natural and anthropogenic aerosol emission in COSMO-ART is an extensive field and was subject to significant changes and further developments in this study. Therefore the emission, as one aspect of the aerosol dynamics, is presented separately in Section 3.1.3 together with the trace gas emissions.

Gas-Phase Chemistry

The chemical reactions of the gaseous species are treated by the Regional Acid Deposition Model Version Karlsruhe (RADMKA; Vogel et al., 2009) that is based on the Regional Acid Deposition Model (RADM2; Stockwell et al., 1990). RADMKA has 60 chemical species, considers 172 chemical reactions and includes several improvements compared to RADM2 which are described in Vogel et al. (2009). Regarding aerosols this includes the update of two heterogeneous reactions than take place at the interface between aerosol and gas phase. The first is the heterogeneous hydrolysis of dinitrogen pentoxide N_2O_5 at the surface of aqueous aerosol particles

$$N_2O_5 + H_2O \xrightarrow{aerosol} 2HNO_3, \tag{3.16}$$

implemented by Riemer (2002). This reaction forms HNO_3 and therefore influences the NO_x budget of the atmosphere. The reaction is linked to the prognostic aerosol of COSMO-ART via their surface density and chemical composition.

The second heterogeneous reaction is related to the formation of nitrous acid HONO (Harris et al., 1982)

$$2NO_2 + H_2O \rightarrow HONO + HNO_3. \tag{3.17}$$

In COSMO-ART, reaction 3.17 is applied for a formation of HONO at aerosol surfaces and at the ground (Vogel et al., 2003).

Photochemistry is affected by the direct and diffuse solar radiation and is therefore also influenced by the distribution of clouds and aerosols. The photochemistry depends on the photolysis frequency j_i that describes the photolysis rate of the species i

$$j_i(\lambda,\tau) = \int F_a(\lambda,\tau)\, \sigma(\lambda,T,P)\, \Phi(\lambda,T,P)\, d\lambda, \tag{3.18}$$

where $F_a(\lambda,\tau)$ is the shortwave actinic flux (light intensity within a certain wavelength interval), σ the absorption cross section of species i and Φ the quantum yield of the photodissociation products, depending on the wavelength λ, AOD τ, temperature T and pressure P. To reduce computing time, Vogel et al. (2009) implemented a parameterization for the photolysis frequencies, combining a look-up table of a priori calculated vertical profiles of shortwave actinic fluxes with the online calculated values for the shortwave actinic flux of the COSMO General Radiative Algorithm Adapted to Linear-type Solutions radiation scheme GRAALS (Ritter and Geleyn, 1992). The photolysis frequency calculations are realized in the photolysis model PAPA (Parameterization of Photolysis Frequencies for Atmospheric Modeling, Bangert, 2007).
The current status of RADMKA does not take into account wet phase chemistry and therefore the sulfate formation in cloud droplets is not taken into account. This can lead to an underestimation of the sulfate burden (Vogel et al., 2009). The deposition of trace gases at the ground is parameterized after Baer and Nester (1993).
The emission characterization of natural and anthropogenic reactive trace gases is described together with the aerosol emission in the subsequent section.

3.1.3 Natural and Anthropogenic Emission Characterization in COSMO-ART

Within the aerosol dynamics and in general for the atmospheric composition, the emission of different pollutants play a pivotal role. Therefore a separate section is dedicated to the representation of the emission in COSMO-ART. This is also motivated to emphasize the significant further developments and updates related to this topic that are realized in this study. The largest contributions are linked to the emission parameterization of mineral dust, biogenic species and gas flaring. Minor contributions are linked to DMS, biomass burning and anthropogenic pollutants. The emission parameterization of sea salt was applied unchanged. In the following, the emission parameterizations are introduced by highlighting the changes realized for this study.

The New Comprehensive Mineral Dust Parameterization

Within this study, the mineral dust emission parameterization of Vogel et al. (2006), that is implemented in COSMO-ART, was extended. In addition to the application of this new comprehensive mineral dust parameterization in COSMO-ART, it was also applied to ICON-ART (Rieger et al., 2017). The new scheme combines Vogel et al. (2006) with the soil disaggregation mechanism of Shao (2004) by using the soil particle size distributions (SPSDs) of Shao et al. (2010) and the global dataset of soil types *Harmonized World Soil Database* (HWSD, 2012). Both, the SPSD and the HWSD data are based on the United States Department of Agriculture (USDA) soil types. The HWSD dataset includes the fraction of USDA soil types with a grid mesh size of 0.0083°. Figure 3.1 shows a map of the predominant HWSD soil types of Africa. The SPSDs of Shao et al. (2010) include the size distributions for the corresponding soil types (see Tab. 3.2). In some cases Shao et al. (2010) uses the same SPSDs for several soil types due to shortage of data as denoted in Table 3.2. To avoid a loss of information when using a coarser mesh size than the original HWSD data, a method to derive fractional soil type coverages per grid cell was applied by Rieger et al. (2017). With these fractional soil types, the dust emission in a grid cell is derived for every soil type fraction separately and finally integrated over all soil types available in the grid cell.

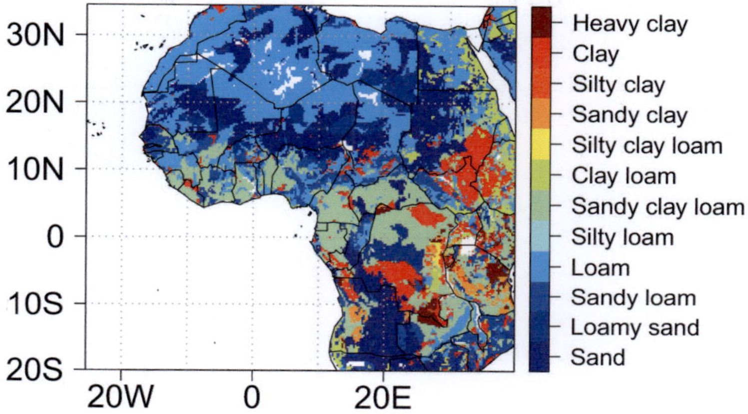

Figure 3.1: Soil types of Africa based on HWSD (2012). Every grid box shows the soil type with the highest fraction in that area.

The combination of HWSD and SPSD of Shao et al. (2010) allows a globally consistent application of the emission parameterization. For every soil type Shao et al. (2010) provides two size distributions, describing the lower and upper limit the SPSD can reach, depending on the friction velocity $u_{\star}$. In case of a low $u_{\star}$, soil aggregates are conserved and the SPSD is shifted to bigger particle diameters. In case of a high $u_{\star}$, the soil aggregates are dispersed and the SPSD is shifted to smaller diameters. Every SPSD of Shao et al. (2010) is given as a sum of two to four lognormal distributions. COSMO-ART also describes the mineral dust with (three) lognormal modes whose corresponding mean diameters for the mass and number distribution as well as the standard deviations are presented in Table 3.1. The parameterization is primarily based on the following eight equations (3.19-3.26), whereas Equation 3.21, 3.23 and 3.24 are included in addition to the preexisting dust emission scheme of Vogel et al. (2006).

Table 3.2: HWSD soil types (HWSD, 2012) and the allocated SPSD (Shao et al., 2010) as well as the clay content based on the USDA soil texture triangle.

HWSD soil type	**SPSD**	**Clay content** f_c (%)
Sand	Sand	5
Loamy sand	Loamy sand	5
Sandy loam	Sandy loam	10
Loam	Loam	15
Silty loam	Silty loam	10
Sandy clay loam	Sandy clay loam (equal to silty loam)	30
Clay loam	Clay loam (equal to loam)	30
Silty clay loam	Silty clay loam (equal to loam)	30
Sandy clay	Sandy clay	45
Silty clay	Silty clay	50
Clay	Clay (equal to silty clay)	80
Heavy clay	heavy clay (equal to silty clay)	100
Silt	Silty loam[a]	5

[a] Silt has no corresponding SPSD, silty loam is used instead.

Equation 3.19 gives the threshold friction velocity for an ideal soil that is dry, non-aggregated and without a salt content (Shao and Lu, 2000)

$$u_{\star t0}(d_p) = \sqrt{A_N(\frac{\rho_p}{\rho_a}\, g\, d_p\, +\, \frac{\gamma}{\rho_a\, d_p})}, \tag{3.19}$$

with the constant values $A_N = 0.0123$ and $\gamma = 3 \cdot 10^{-4}\ kg\ s^{-2}$, the particle density $\rho_p = 2650\ kg\ m^{-3}$ and air density ρ_a, the soil particle diameter d_p and the gravitational acceleration g. By using correction functions for soil moisture f_ξ and roughness f_r, the friction velocity for an ideal soil $u_{\star t0}(d_p)$ (Eq. 3.19) is transformed to $u_{\star t}(d_p)$ for a real soil (that is less erosive in case of higher soil moisture and or higher roughness length) via Equation 3.20.

$$u_{\star t}(d_p) = u_{\star t0}(d_p)\ f_r\ f_\xi \tag{3.20}$$

The correction factor according to roughness f_r is determined by Raupach et al. (1993)

$$f_r = \sqrt{1 - 0.5\ \lambda'}\ \sqrt{1 + 0.5\ 90\ \lambda'};\ \lambda' = -0.35\ ln(f_b), \tag{3.21}$$

where f_b is the fraction of bare soil. For the present study the bare soil fraction is estimated as $f_b = c_b + 0.5\ c_s$ by using the landuse classes bare soil c_b and sparse vegetation c_s of the GlobCover 2009 dataset (GlobCover, 2009). The correction factor according to soil moisture follows Fécan et al. (1999)

$$f_\xi = \sqrt{1 + 1.21 * (\xi - \xi')^{0.68}} \text{ with} \tag{3.22}$$

$$\xi' = a_z\ (0.0014\ f_c^2 + 0.17\ f_c),$$

where ξ (%) is the gravimetric soil moisture and ξ' the minimum gravimetric soil moisture depending on the clay content f_c as given in Table 3.2. The parameter $a_z = 5$ is a factor to account for too high soil moisture in the model (Rieger et al., 2017). For the mineral dust emission the soil-atmosphere interface is pivotal and this top soil layer can be linked to drying that is likely not or insufficiently represented in the model (Deetz et al., 2016). The correction functions f_r and f_ξ are 1 in case of negligible soil moisture and roughness and increase with increasing influence of soil moisture and roughness. For a friction velocity $u_\star$ greater than $u_{\star t}$ an emission of mineral dust takes place.

As described above, for every soil type two SPSDs were provided by Shao et al. (2010), one size distribution for a soil with conserved soil aggregates $p_m(d_p)$ (minimal dispersed) and one for a soil without aggregates $p_f(d_p)$ (fully dispersed). The SPSD $p(d_p)$ related to the current wind conditions is derived via Equation 3.23

$$p(d_p) = \gamma_d\ p_m(d_p) + (1 - \gamma_d)\ p_f(d_p), \tag{3.23}$$

where γ_d (between 0 and 1) is the disaggregation factor describing the location of the realistic SPSD depending on $u_\star$ (Shao, 2004)

$$\gamma_d = exp(-0.5(u_\star - u_{\star t,min})^3), \tag{3.24}$$

where $u_{\star t,min}$, the minimum of $u_{\star t}$, is given by the first derivative of Equation 3.19. The saltation flux is given by White (1979)

$$F_h(d_p) = c_{white}\ \frac{\rho_a}{g}\ u_\star^3 \left(1 + \frac{u_\star(d_p)}{u_\star}\right) \left(1 - \frac{u_\star^2(d_p)}{u_\star^2}\right), \tag{3.25}$$

with the tuning parameter c_{white} that is estimated as 0.75 based on COSMO-ART realizations and comparison with MODIS deep blue AOD observations in the Saharan area. Finally, the dust emission is calculated by Equation 3.26 (Alfaro and Gomes, 2001)

$$F_{v,l}(d_p) = \frac{\pi}{6}\ \rho_p\ d_l^3\ \frac{p_l(d_p)\ F_{kin}(d_p)}{e_l}, \tag{3.26}$$

where $p_l(d_p)$ are the partitioning functions for the kinetic energy to the lognormal dust modes $l = 1,2,3$, the kinetic energy flux $F_{kin} = 163\ m\ s^{-2}\ F_h(d_p)$ and the particle binding energy e_l. The mineral dust emission is calculated for the potential emission area (snow, ice, rock, water and urban areas are excluded).

Sea Salt and Dimethyl Sulfide

The sea salt is described with three lognormal modes for the film, jet and spume droplets, respectively (see Tab. 3.1). The aerosols can consist of sodium chloride, sodium sulfate and sulfuric acid. The film droplets (D_p=0.02-1 μm) are parameterized by Mårtensson et al. (2003), the jet droplets (D_p=1-9 μm) by Monahan et al. (1986) and the spume droplets (D_p=9-28 μm) by Smith et al. (1993). All three parameterizations depend on the 10 m horizontal wind speed. The sea salt emission from film droplets additionally depends on the sea water temperature. No changes were made for the sea salt representation in COSMO-ART for this study.
DMS emission is realized by phytoplankton or bacterial transformation of dimethyl sulfoxide (DMSO). The DMS fluxes were read from an external dataset. For this study the DMS flux climatology of Kettle et al. (1999) was replaced by the updated dataset of Lana et al. (2011). The actual emission of DMS in COSMO-ART is a function of the DMS flux and the 10 m horizontal wind speed.

The New Representation of Biogenic Volatile Organic Compounds (BVOCs)

To consider the emissions from terrestrial ecosystems, the existing biogenic emission routine in COSMO-ART was replaced in this study by a state-of-the-art parameterization based on the *Model of Emissions of Gases and Aerosols from Nature version 2.1* (MEGAN2.1) (Guenther et al., 2012). In Guenther et al. (2012) the parameterization is described in detail, therefore this subsection only describes the changes and adaptions that were made to implement the parameterization into COSMO-ART and to make it applicable for the conditions of SWA. The resulting BVOC emission module also became

a model part of ICON-ART and was successfully applied for global simulations of acetone concentrations (Weimer et al., 2017).
The representation of BVOCs in atmospheric chemistry modeling is highly relevant since the BVOCs are an important precursor in the formation of ozone (e.g. Monks et al., 2015). In COSMO-ART the parameterization considers temperature, photosynthetically active radiation, leaf area index (LAI), plant functional types (PFTs), leaf age and leaf angle. The dependency with respect to soil moisture and carbon dioxide concentration was not implemented since it is problematic to allocate a modeled soil moisture to the different PFTs. Furthermore, COSMO-ART does not consider CO_2 and therefore cannot capture the annual CO_2 cycle.
Temperature and photosynthetically active radiation are provided by the meteorological part of the model. The LAI is given by the external parameter dataset of COSMO and the PFTs are provided from the external dataset of Lawrence and Chase (2007) using MODIS remote sensing. The data can be obtained from the Community Climate System Model (CCSM, 2015) with a grid mesh size of 0.05°. Figure 3.2 shows that the predominant five PFTs for SWA are: broadleaf evergreen tropical tree (Fig. 3.2a), broadleaf deciduous tropical tree (Fig. 3.2b), cool C3 grass (Fig. 3.2c), warm C4 grass (Fig. 3.2d) and crop (Fig. 3.2e).
For the leaf age a spatiotemporally uniform distibution between the classes new, growing, mature and senescing foliage is assumed, because COSMO-ART does not include a biome model. In addition, we added the parameterization of Dai et al. (2004) to estimate the fraction of the LAI that is lit by sun since only this LAI part can emit BVOCs. By using Dai et al. (2004), an overestimation of BVOC emissions especially in tropical evergreen forests with high LAI and several vegetation layers is avoided. The parameterization needs the leaf angle as an input parameter. A spatiotemporal constant value of 0.8 for all PFTs (1 is horizontal, 0 spherical and -1 vertical) was used, together with a constant sun zenith angle of 10.4°. The biogenic emission F ($\mu g\ m^{-2}\ h^{-1}$) of a compound i is finally calculated via Equation 3.27

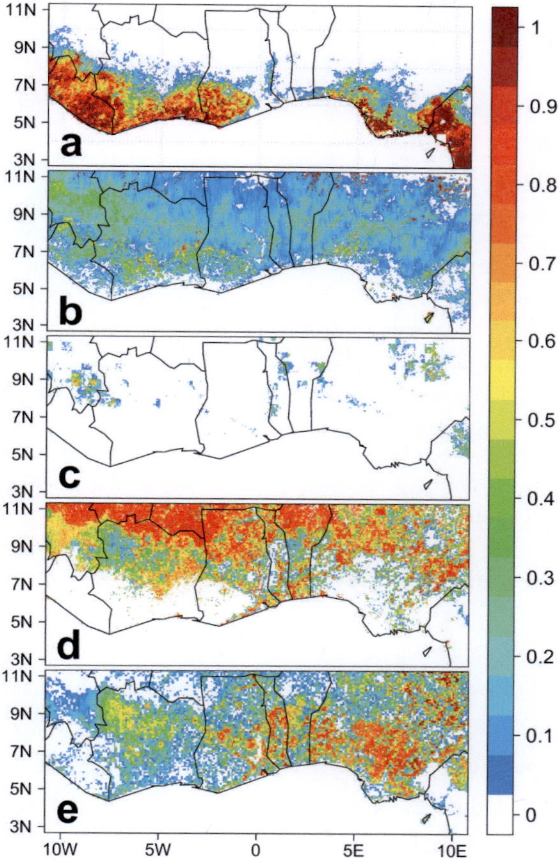

Figure 3.2: Fractional coverage of PFTs in SWA (CCSM, 2015): (a) broadleaf evergreen tropical tree, (b) broadleaf deciduous tropical tree, (c) cool C3 grass, (d) warm C4 grass and (e) crop.

$$F_i = \gamma_i \sum \varepsilon_{i,j} \chi_j, \tag{3.27}$$

with the activity factor γ for the different processes controlling the emissions, in particular radiation, temperature, leaf area index and leaf age. The emission factor ε depends on the compound class i and the PFT j. χ is the fractional coverage of the PFT j in a grid cell. Guenther et al. (2012) delivers 19 classes of biogenic compounds. COSMO-ART makes use of three of these classes: isoprene, α-pinene and limonene. Table 3.3 summarizes the emission factors for the five relevant PFTs for SWA (Fig. 3.2) by focusing on the three BVOC species considered in COSMO-ART.

Table 3.3: Emission factors ε (μg m^{-2} h^{-1}) of the COSMO-ART BVOCs for the relevant PFTs in SWA after Guenther et al. (2012).

Compound class	Evergreen tropical	Deciduous tropical	C3 grass	C4 grass	Crop
Isoprene	7000	7000	800	200	1
α-Pinene	600	600	2	2	2
Limonene	80	80	0.7	0.7	0.7

Table 3.3 indicates that the highest biogenic emissions can be expected from isoprene in the broadleaf evergreen tropical (Fig. 3.2a) and broadleaf deciduous tropical (Fig. 3.2b) areas. Broadleaf deciduous tropical forests are more widespread than the broadleaf evergreen tropical forests but with a generally lower fractional coverage. Therefore the highest emissions are linked to Liberia, southern Ivory Coast, southern Ghana and southeastern Nigeria. For α-pinene and limonene the emissions in these areas are one and two orders of magnitude smaller, respectively. Cool C3 grass also shows high isoprene emissions but the coverage in SWA is very low (Fig. 3.2c). For warm C4 grass (Fig. 3.2d) the emissions are significant lower compared to the tropical forest but a high coverage can be observed in the entire northern part of SWA. Crop shows lowest emissions but has a high coverage in the central latitudes of SWA. In general, a dominance of the emission of isoprene and α-pinene in

the southern part, α-pinene in the central part and isoprene in the northern part of SWA can be expected. However, it has to be considered that isoprene emissions depend on the availability of the photosynthetically active radiation in contrast to α-pinene and limonene.

Biomass Burning Emissions with Vegetation Type-Dependent Diurnal Cycles

Biomass burning occurs only sporadically in the research domain in boreal summer but is predominant in central Africa, whereas in boreal winter the biomass burning pattern is shifted to higher frequencies in SWA and nearly no fires in central Africa. These patterns can be quantified via remote sensing data from the Global Fire Assimilation System (GFASv1.2) (Kaiser et al., 2012) that can be downloaded as daily global data via the CAMS Global Fire Assimilation System (CAMS, 2017a). The mean frequency of occurence of wild fires in boreal winter (December 2015 - February 2016) and for the DACCIWA campaign period (June-July 2016) is shown in Figure 3.3a and 3.3b, respectively. Among the fire location, this product uses the fire radiative power (FRP in W m^{-2}) derived by remote sensing techniques based on data of MODIS Terra and Aqua and links this measure to the dry matter combustion rate to derive biomass burning emissions.

Fire frequencies in boreal summer in a range of 40-50% can be observed in the southern part of the Dem. Rep. of the Congo, Angola and Zambia (Fig. 3.3b). This illustrates that on average every second day in the time period June-July 2016 a fire was detected by satellite in that region. The GFAS product delivers the FRP as well as a list of emission species of gaseous and aerosol compounds. Figure 3.4 shows (a) the mean FRP and (b) the mean CO emissions in June-July 2016. By comparing Figure 3.3 with Figure 3.4a and 3.4b, it is obvious that the area with the highest fire frequency of occurrence coincide with the area of most intense fires regarding the FRP and therefore also the amount of emis-

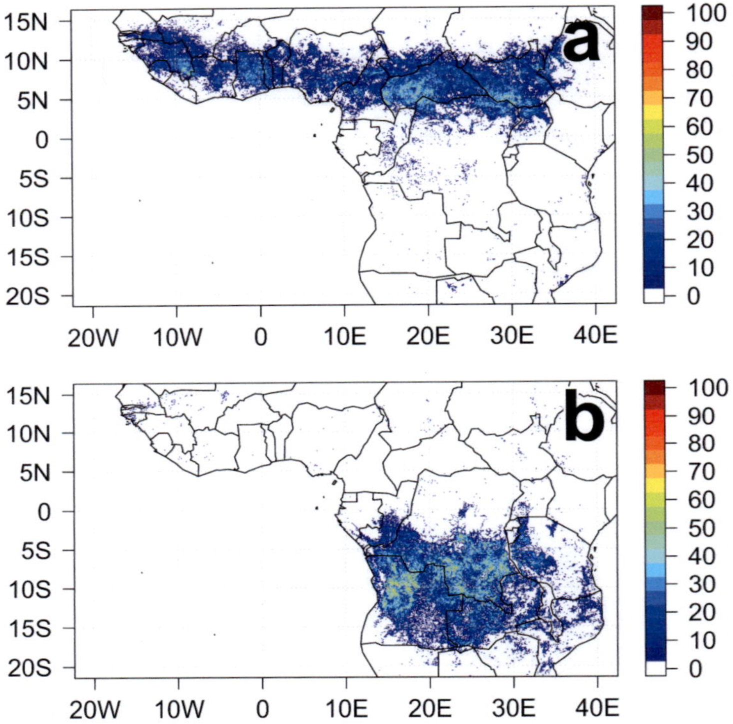

Figure 3.3: GFAS wild fire frequency of occurence (%) within the time period (a) December 2015 to February 2016 and (b) June 2016 to July 2016.

sions. Despite the spatial distance between the research area and the biomass burning hotspots of about 2500 km in boreal summer, the biomass burning emissions have to be considered in the simulations since a long-range transport of the polluted air in northwesterly direction to the area of interest frequently occurs (e.g. Mari et al., 2008). This underlines the need for an enlarged modeling domain in southeasterly direction to represent these emissions in the model realizations.

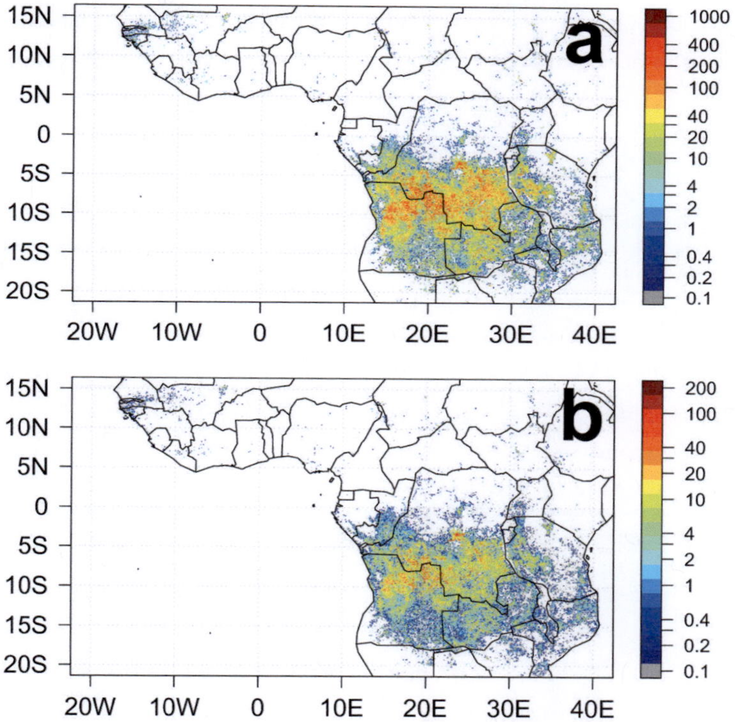

Figure 3.4: GFAS (a) fire radiative power (FRP, mW m^{-2}) and (b) CO emissions (mg m^{-2} h^{-1}), mean over June-July 2016

Walter et al. (2016) implemented the one-dimensional subgrid-scale plume rise model of Freitas et al. (2006, 2007, 2010) into COSMO-ART to enable the model to read the GFAS emission data and to derive the injection height of the pollution plume, assuming a constant fire area of 0.5 km^{-2}. The minimum (top) level of the injection layer is derived by using the lower (upper) limit heat flux for tropical forests (see Tab. 3.4). Walter et al. (2016) also implemented a function to transform the daily values of fire size, fire intensity and emissions to a diurnal cycle to consider the findings of Zhang and Kondragunta (2008) in which the peak fire activity (PFA) occurs between 10 and 15 local solar time due to a reduced moisture and enhanced wind speed that strengthen the fire intensity.

Table 3.4: Heat flux lower (H_{min}) and upper (H_{max}) limits for different vegetation types (V1-V3) according to Freitas et al. (2006) and the fraction of daily biomass burning emissions (f_e) that are emitted during peak fire activity (PFA, 10-15 local solar time) according to Zhang and Kondragunta (2008).

Vegetation type (VT)	H_{min} $(kW\ m^{-2})$	H_{max} $(kW\ m^{-2})$	f_e (%)
(1) Tropical forest	30.0	80.0	52.1
(2) Woody savanna - cerrado	4.4	23.0	61.8
(3) Grassland - pasture - cropland	3.3	3.3[b]	73.7[c]

[b] For this vegetation type, Freitas et al. (2006) does not provide an upper limit value. In the following, the same as for the lower limit is used.
[c] By taking the average between grassland (62.5%) and cropland (84.9%).

They found that the fraction of daily emissions that are released during PFA (f_e hereafter) is 52.1% for tropical forest. Walter et al. (2016) implemented this diurnal cycle for forests in COSMO-ART. This is a valid approach for the study of Canadian forest fires presented in Walter et al. (2016) but is problematic when using the model in different areas with different vegetation types. For this study we therefore extended the diurnal cycle consideration also to savanna and grassland. Table 3.4 summarizes the f_e for different vegetation types. Since only the heat flux values for the three vegetation types presented in Table 3.4 are available, the 23 landuse classes of GlobCover (2009) applied in COSMO-ART are assigned to these three types (see Tab. 3.5).
Two diurnal cycles are defined in the present study, the first for describing the heat flux (see Fig. 3.5a) and the second for describing the emissions and the burned area (see Fig. 3.5b). For the first diurnal cycle the two conditions:

(a) all curves reach the value 1 within the PFA to ensure that the maximum heat flux is reached at that time and (b) f_e is reached within the PFA (see gray area in Fig. 3.5a) are fulfilled.

Table 3.5: Assignment of the GlobCover2009 landuse classes to the vegetation types (VT) specified in Table. 3.4. The landuse classes *artificial surfaces, bare areas, water bodies, permanent snow and ice* and *undefined* are not considered.

GlobCover landuse classes	VT1	VT2	VT3
Irrigated croplands			•
Rainfed croplands			•
Mosaic cropland (50-70%) - vegetation (20-50%)			•
Mosaic vegetation (50-70%) - cropland (20-50%)			•
Closed broadleaved evergreen forest	•		
Closed broadleaved deciduous forest	•		
Open broadleaved deciduous forest	•		
Closed needleleaved evergreen forest	•		
Open needleleaved deciduous or evergreen forest	•		
Mixed broadleaved and needleleaved forest	•		
Mosaic shrubland (50-70%) - grassland (20-50%)			•
Mosaic grassland (50-70%) - shrubland (20-50%)			•
Closed to open shrubland		•	
Closed to open herbaceous vegetation			•
Sparse vegetation			•
Closed to open forest regularly flooded	•		
Closed forest or shrubland permanently flooded	•		
Closed to open grassland regularly flooded			•

For the second diurnal cycle the two conditions: (a) the integral of the curves are 1 to ensure that in every case the same amount of emissions and the same fire area is distributed over the day and (b) f_e is reached within the

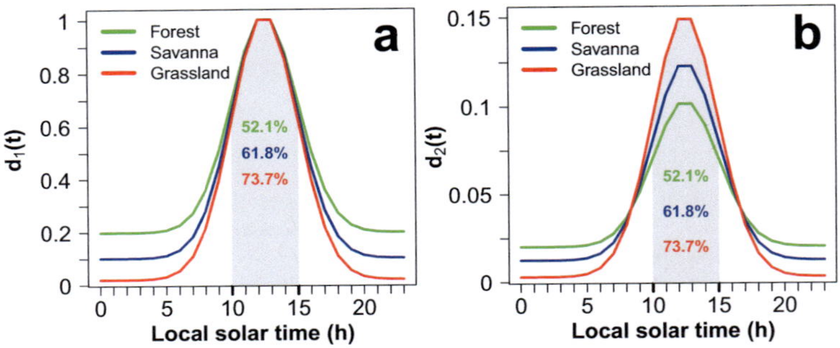

Figure 3.5: Vegetation type-dependent general diurnal cycle for (a) the heat flux and (b) the fire emissions and the fire area. The gray shaded area denotes the peak fire activity (PFA) between 10 and 15 local solar time and the percentages describe the vegetation type-dependent fractions of daily emissions that are released during the PFA.

PFA (see gray area in Fig. 3.5b) are fulfilled. Following Walter et al. (2016) also for this study a weighted Gaussian distribution was used to describe the diurnal cycle, but with new weightings (see Tab. 3.6) to satisfy the above-mentioned conditions.

Table 3.6: Weightings for the Gaussian distribution describing the vegetation type-dependent diurnal cycle of the heat flux limits, the emissions and the fire area.

Vegetation type (VT)	ω_1	ω_2
Tropical forest	0.03870	9.890
Woody savanna - cerrado	0.01754	8.170
Grassland - pasture - cropland	0.00308	6.735

Equation 3.28 shows the Gaussian distribution to describe the diurnal cycle

$$d(t) = \omega_1 \, \frac{1}{\sigma^* \sqrt{2\pi}} \, exp\left(-0.5\left(\frac{t - t_0{}^2}{\sigma^*}\right)\right), \tag{3.28}$$

with the expected value $t_0 = 12.5$ h, the standard deviation $\sigma^* = 2.5$, the local solar time $t(h)$ (Walter et al., 2016) and the weighting ω_1 (see Tab. 3.6), which ensures that f_e is reached within the PFA. For d_2, related to the emissions and the burned area (see Fig. 3.5b), $d(t)$ is normalized

$$d_2(t) = \frac{d(t)}{\sum d(t)}. \tag{3.29}$$

For d_1, related to the heat flux (see Fig. 3.5a), d_2 is multiplied with the weighting ω_2 (see Tab. 3.6) to ensure that all curves are reaching the value 1 (maximum heat flux) in the PFA.

$$d_1(t) = \omega_2\, d_2(t) \tag{3.30}$$

Fig. 3.6 shows an example for a grid cell containing one third tropical forest, one third savanna and one third grassland. For the red curve of the heat flux lower limit (Fig. 3.6a), every distribution $d_1(t)$ for the different vegetation types are multiplied with their fractional coverage in the grid cell and their corresponding lower limit values H_{min} of Table 3.4. Finally, the three distributions are summed up. The same is done for the blue curve but multiplying with H_{max}. By using the red and the blue curve, the diurnal cycle of the lower and upper limit of the injection layer can be derived.

To derive the diurnal cycle for the emissions (see brown curve in Fig. 3.6b), every distribution $d_2(t)$ for the different vegetation types are multiplied with their fractional coverage within the grid cell and the emission flux integrated over one day (e.g. 14400 kg m^{-2} d^{-1} in Fig. 3.6b). For the fire area (burned area), $d_2(t)$ of the different vegetation types are multiplied with the fire area of 0.5 km^2 (see black solid curve in Fig. 3.6b). In addition, the dashed black curve of Figure 3.6b shows the accumulated fire area, which is again 0.5 km^2 at the end of the day, as desired.

For this study, we had access to near real-time daily data of GFAS fire observations (one day offset) with a grid mesh size of 0.1°.

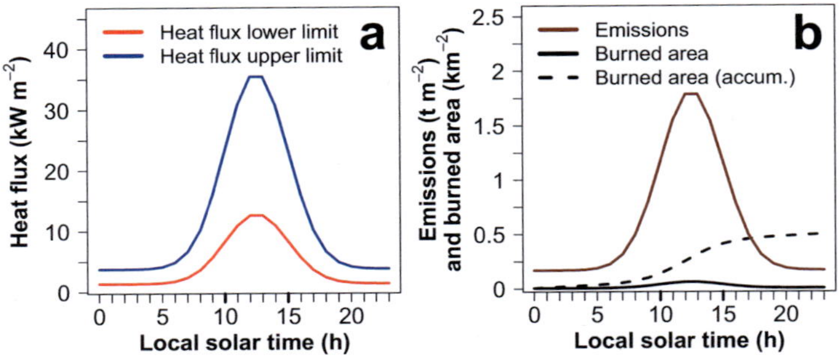

Figure 3.6: Exemplary diurnal cycle for a grid cell containing one third tropical forest, one third savanna and one third grassland regarding (a) the heat flux lower limit (red) and upper limit (blue) (kW m^{-2}) and (b) the emissions (brown, t m^{-2}) as well as the burned area (black solid, km^{-2}). The dashed black line shows the accumulated burned area.

The data was preprocessed for the use in COSMO-ART to assign the GFAS species to the species defined in COSMO-ART. Among the FRP, emissions of sulfur dioxide, carbon dioxide, carbon monoxide, alkanes, alkenes, black carbon, isoprene, α-pinene, formaldehyde, acetaldehyde, ketones, ammonia, DMS, xylene, nitrous oxide, nitrous dioxide, PM2.5 and PM10 (kg m^{-2} s^{-1}) were used. No scaling factor was used to increase or decrease the GFAS emissions.

Nitrous Oxide Emissions from Soil

The emissions of nitrous oxides from soil (soil NO_x) is parameterized in COSMO-ART by Ludwig et al. (2001). For this study the soil NO_x emissions are not considered due to two reasons: 1. There is a substantial shortcoming in detailed and up-to-date data for landuse, agriculture and soil fertilization that are important information to estimate the soil NO_x. 2. For example Vinken et al. (2014) showed by using remote sensing observations of NO_2 from the Ozone Monitoring Instrument (OMI) for 2005 that emissions of soil NO_x

are substantial in the Sahel approximately between 10°N and 15°N with 4-10 ng N m^{-2} s^{-1} (highest values between April and June) but not in the area of interest south of 10°N with values below 3 ng N m^{-2} s^{-1}. Also observations from AMMA confirm the Sahel as the dominant emission area of soil NO_x (Mari et al., 2011). With the focus on June and July, that is linked with a predominant northeasterly flow, it is assumed that the influence of soil NO_x on the area of interest is negligible.

Anthropogenic Emissions

The anthropogenic emissions are represented in COSMO-ART via pre-calculated external datasets. For this study the EDGAR HTAP_v2 dataset (EDGAR, 2010) was used. The dataset delivers monthly data for 2010 with a grid mesh size of 0.1° including the emission sectors air traffic, ship traffic, energy, industry, transport, residential and agriculture. Emissions of the gases carbon monoxide, ammonia, nitrogen oxide (splitted in 10% nitrogen dioxide and 90% nitrogen monoxide), sulfur dioxide, soot, PM10, PM2.5 (splitted in 10% to Aitken mode and 90% to accumulation mode), organic carbon (splitted in 10% to Aitken mode and 90% to accumulation mode), ethene and the VOCs: alkanes, alkenes, toluene, xylene, cresol, ketone, aldehyde, formaldehyde and acids. The VOC speciation was used from ECCAD2.
Additionally, a diurnal cycle was implemented in this study for all emission sectors except air traffic. Due to missing information about the diurnal cycle of anthropogenic emissions in the research area, a diurnal cycle for Morocco for Monday, 2 June 2008 from the Transport and the Environment (VEU, 2017) dataset was used for all simulations without considering a weekly cycle. Figure 3.7 shows the diurnal cycle for CO, whereupon the differences between the emission species are negligible.

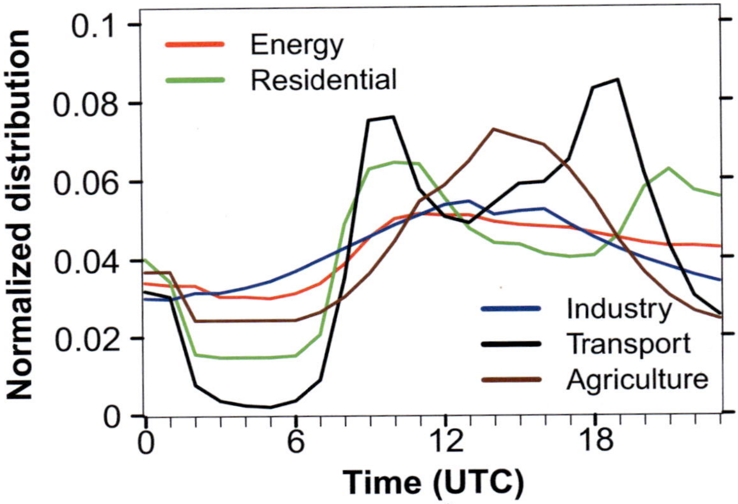

Figure 3.7: Normalized diurnal cycle of CO emission for the emission sectors: energy (red), residential (green), industry (blue), transport (black) and agriculture (brown). The diurnal cycle is taken from VEU dataset for Morocco (Monday, 2 June 2008). Emissions from air traffic are not linked to a diurnal cycle.

A New Method to Derive Gas Flaring Emissions

This study for the first time provides a method to derive point source trace gas emissions from gas flaring via the combination of remote sensing observations and physically-based combustion equations. Due to the outstanding relevance of this development, which is published in Deetz and Vogel (2017), the gas flaring representation is described in this separate section, although gas flaring is a part of the anthropogenic emissions. In the following, the main idea, the methods and the compilation of the dataset for the use in COSMO-ART are presented. Based on this method, as a byproduct, a daily gas flaring emission inventory (GFEI hereafter), capturing the time period March 2012 to August 2016 and including six inventories with different assumptions, was derived and made available for the DACCIWA research community. This dataset is described in the final part of the section.

Motivation for the Model Representation of Gas Flaring

Gas flaring is a globally used method to dispose flammable, toxic or corrosive vapors to less reactive compounds at oil production sites and refineries. In regions of insufficient transportation infrastructure or missing consumers, flaring is also commonly applied. The implications of gas flaring are far-reaching by influencing the environment by noise and deterioration of the air quality (Osuji and Avwiri, 2005). Higher concentrations of heavy metals in surface water of a gas flared environment in Delta State Nigeria are measured by Nwankwo and Ogagarue (2011). Nwaugo et al. (2006) indicate adverse ecolological and bacterial spectrum modifications by gas flaring. Further implications of gas flaring are acid rain, which causes economic burden via rapid corrosion of zinc roofs (Ekpoh and Obia, 2010) and the retardation in crop growth due to high temperatures during the combustion (Dung et al., 2008). The emissions usually include soot, unburned hydrocarbons, CO, CO_2, partially burned and altered hydrocarbons as well as NO, NO_2 and SO_2 (EPA, 2015).

The research area SWA is affected by emissions from gas flaring. Especially Nigeria contributes to the flaring pollution in SWA. Nigeria is the second largest flaring country (15 bcm flared gas in 2012) after Russia (35 bcm flared gas in 2012) (Elvidge et al., 2016).

Therefore importance is attached to consider these emissions in the COSMO-ART simulations. Although gas flaring is already included in the industrial sector of the anthropogenic emission dataset EDGAR HTAP_v2 (Greet Maenhout, personal communication), we decided to develop and apply a new inventory to consider this kind of emission. The main reason was that the HTAP_v2 gas flaring emissions were smeared out with the rural population proxy data due to shortcoming in spatial flaring data. To smear out the flaring with the population data seems to be highly problematic, because flaring sites are characteristic point sources and these flaring sites are usually located away from populated areas or are sufficiently isolated (EPA, 2015). Additionally, this approach neglects all offshore flares and has an offset of six years to the research period June-July 2016. A better representation will be given in the further developed inventory EDGARv4.3.2 but this inventory was not available when performing this study. The drawback of adding a further emission inventory is the potential double count of flaring emissions in certain areas since it was not possible to exclude the flaring emissions from HTAP_v2. We assume that the effect of the double count is small since the majority of the flaring sites are offshore. The Niger Delta is the only area where a double count of flaring emissions can be relevant.

To describe the emissions from gas flaring more realistic, a new dataset was developed. The parameterization to derive the emissions is described in detail by Deetz and Vogel (2017), including an assessment of the various sources of uncertainty. Therefore in this section only the main idea and the result of this study is presented.

Parameterization of Gas Flaring

The new flaring parameterization is based on the remote sensing product *VIIRS Nightfire Nighttime Detection and Characterization of Combustion Sources* (VNF hereafter) (VIIRS, 2016) and the combustion equations of Ismail and Umukoro (2014) (IU14 hereafter). VIIRS stands for *Visible Infrared Imaging Radiometer Suite*. VNF is available as daily datasets including the flare locations, source temperatures, radiant heat and time of observation. The VNF dataset was preprocessed by the Earth Observation Group of the National Oceanic and Atmospheric Administration NOAA (VNF_{flare}) to select the countries that can affect SWA with their flaring emissions (in particular Ivory Coast, Ghana, Nigeria, Cameroon, Gabon, Congo, the Democratic Republic of the Congo and Angola) and to isolate the flaring sites from other combustion

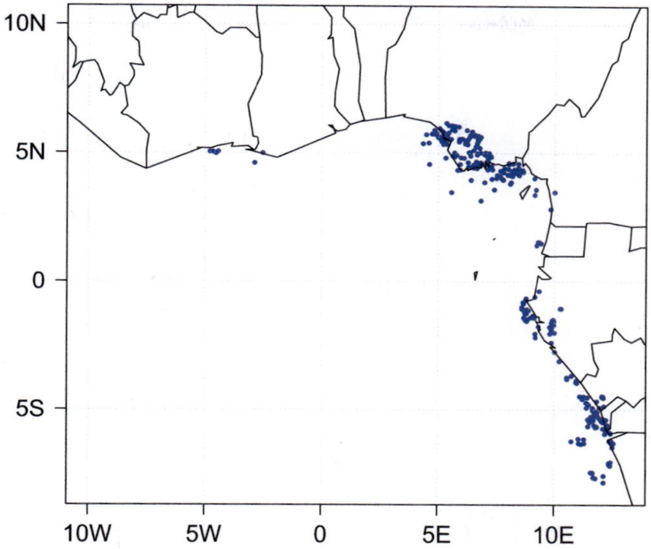

Figure 3.8: Location of the 227 flares relevant for SWA regarding June-July 2016

sources (e.g. wild fires). Three time periods are considered for the flaring emissions: a) June-July 2015 for the COSMO-ART forecasts (since at this time the data for 2016 were not available), b) June-July 2016 for post DACCIWA campaign case studies (available since 23 November 2016) and c) March 2012 to August 2016, the full time period (available since 23 November 2016), provided to the DACCIWA database (baobab.sedoo.fr, last access: 10 June 2017) for the DACCIWA research community.

For the modeling studies with COSMO-ART, the focus is located on a two-month time period of observations from 1 June to 31 July 2016 to cover the period of the DACCIWA measurement campaign and to allow for a compilation of a flaring climatology in terms of the flare locations and emissions to avoid a flaring underestimation due to flares masked by clouds. Figure 3.8 shows the location of the 227 relevant flares for SWA in

June-July 2016. The second part of the parameterization are the combustion equations of IU14. The flaring emissions were estimated based on equations for incomplete combustion depending on the availability of sulfur in the flared gas (that determines the SO_2 formation) and the source temperature (that determines the NO and NO_2 formation). As input, IU14 needs the natural gas composition that is taken from Nigerian natural gas flow station measurements (Sonibare and Akeredolu, 2004), the source temperature T_s (K) and radiant heat H (MW) given by VNP, the flare characteristics including combustion efficiency $\eta = 0.8$ and availability of combustion air $\delta = 0.95$ that is estimated by usual values given in IU14, the gauge pressure of the fuel gas in the flare p_f, that is estimated as 5 psi (34.475 kPa) (API, 2007) and finally the fraction of total reaction energy that is radiated $f = 0.27$ (Guigard et al., 2000).
The flow rate F (m^{-3} s^{-1}) can be calculated via Equation 3.31 (VDI, 1985):

$$F = \frac{H_f}{c_p^*(T_s - T_a)}, \quad (3.31)$$

where H_f is the emitted heat flow, which is assumed to be equal to the total reaction energy of the flare. VNF$_{flare}$ only detects the energy fraction that is radiated H and not the total energy H_f. Therefore H_f is estimated by H/f. c_p^* is the mean specific heat capacity of the emissions. VDI (1985) provides a value of the mean specific heat capacity of

$$c_p^* = 1.36 \cdot 10^{-3}\, MW\, s\, m^{-3}\, K^{-1}, \quad (3.32)$$

that is derived for a pit coal firing but VDI (1985) denotes, that this can be used for other flue gases as well since potential deviations are negligible. T_a is the ambient temperature that is estimated as 298.15 K as a fixed value, representative for the tropical region. Within a sensitivity study regarding the influence of T_a on the heat flow, the averaged heat flow and source temperature of all flares within June-July 2015 were used together with a variation of the ambient temperature between 293 K and 303 K, as a reasonable temperature range in the tropical regions. The resulting maximum difference in the heat flow is 0.0036 $m^3\ s^{-1}$. Therefore it is assumed that the uncertainties using a fixed climatological value for the ambient temperature are negligible. By using Equation 3.31 and 3.32 the heat flow F can be derived as

$$F = \frac{H_f}{1.36 \cdot 10^{-3}\,(T_s - 298.15)}. \tag{3.33}$$

This is the key equation since it allows for a synthesis of VNP and IU14. The estimation of the fuel gas density ρ_f is necessary to transform F into an emission of a mass. Via Equation 3.34 ρ_f can be derived as

$$\rho_f = \frac{p_f\, M_f}{R\, T_a}, \tag{3.34}$$

where R is the universal gas constant and M_f the molar mass of the fuel gas. Finally, the emission E (kg s^{-1}) of a species i is given by

$$E_i = \frac{m'_i}{m'_{total}} \rho_f\, F, \tag{3.35}$$

where m_i is the mass of the species i and m_{total} the total mass of the fuel gas, both delivered by the parameterization of IU14. The combustion calculations within IU14 provide the species water, hydrogen, oxygen, nitrogen, carbon monoxide, carbon dioxide, sulfur dioxide, nitrogen oxide and nitrogen dioxide. In the following only the latter five are considered. However, no black carbon and no VOCs are considered by IU14, although they are not negligible (e.g. Johnson et al., 2011; Strosher, 1996).

Gas Flaring Dataset for COSMO-ART

In the preparation for this work, the locations of the flares in June-July 2014 were compared with the Google Earth imagery (GoogleEarth, 2014). Although only the onshore flares are visible in Google Earth, the visual verification

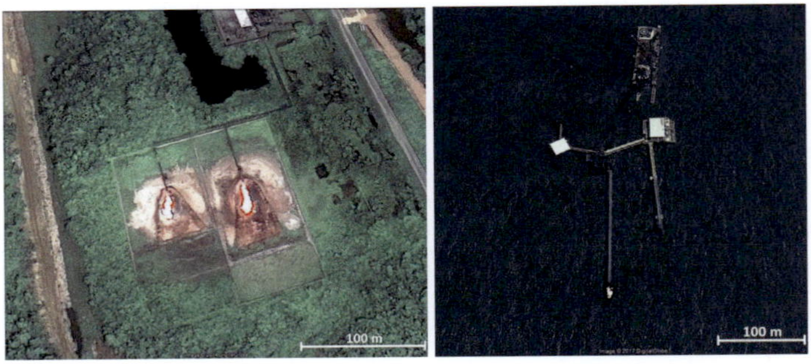

Figure 3.9: Flares in Nigeria observed by GoogleEarth (2014) (Image © 2014 DigitalGlobe) and detected by VNF$_{flare}$ in 2014: (left) an onshore flare in central Niger Delta (4°53'08.04"N 6°22'20.81"E; 2009) and (right) an offshore flare in the western part of the Niger Delta (5°22'36.88"N, 5°18'29.09"E; 2013). A size scale (m) is added in the bottom right corner of the figures.

(not shown) reveals that 72% of the VNF$_{flare}$ detected onshore flares are visible in Google Earth. Often the image quality is not good enough for verification or the images are not up to date, so the hit rate is very likely much higher. This comparison confirms the reliability of the VNF$_{flare}$ method to identify the flares in SWA. Figure 3.9 shows two examples of these flares, one onshore (Fig. 3.9a) and one offshore (Fig. 3.9b) flaring site. Figure 3.10 shows the flaring emission inventory for June-July 2016 for the species CO, SO_2, NO and NO_2. For Figure 3.10 the emissions for all detected flares are calculated separately on a daily basis, using the daily observations of H but monthly averages of T_s because the source temperature detection is sensitive on atmospheric conditions (Mikhail Zhizhin, personal communication). For the flares that are cloud-covered at a specific day, it is assumed that these flares are active but not observed due to cloud masking. For these flares the yearly mean of emission was used at this day. The 61 day flaring dataset was than temporally averaged for every flare. In this case all flares are active at once with their mean emission strength. This dataset is used for the case study realization with COSMO-ART, described in Section 6.

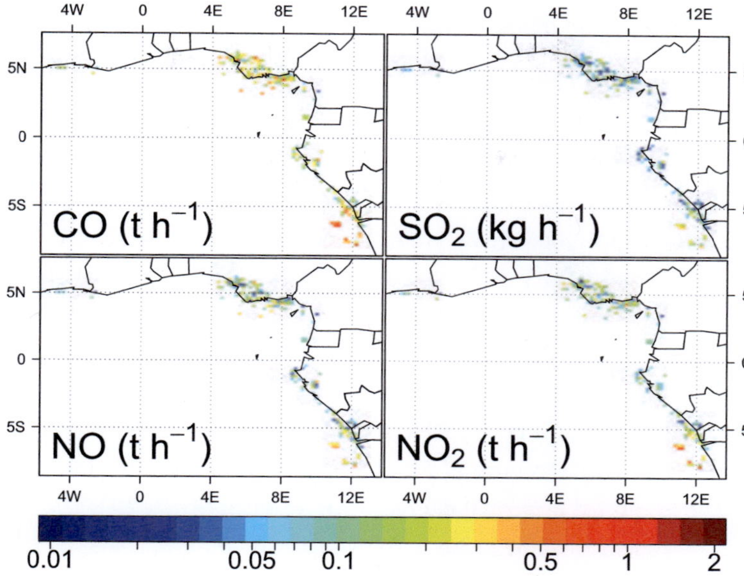

Figure 3.10: Flaring emission inventory for June-July 2016 for CO (t h^{-1}), SO_2 (kg h^{-1}), NO (t h^{-1}) and NO_2 (t h^{-1}), interpolated on a 0.2° grid.

Highest emissions are calculated for CO_2 (not shown), followed by CO, NO_2, NO and SO_2. SO_2 shows lowest emissions since these emission depend only on the natural gas composition and on the amount of flared gas but not on combustion processes. Deetz and Vogel (2017) places the estimated flaring emissions in the context of existing emission inventories regarding CO_2. The main sources of uncertainty addressed by Deetz and Vogel (2017) are due to the uncertainty in the IU14 parameters, the gauge pressure of the flare and the fraction of total reaction energy that is radiated. The flaring emissions presented in Figure 3.10 are saved in an ASCII file and read as point sources into COSMO-ART. Based on the source temperature, the plume superelevation is calculated to find the correct injection height in the model.

Gas Flaring Emission Inventory (GFEI)

As a byproduct, the GFEI was derived, which is a daily gas flaring emission dataset based on the method described above and capturing the time period March 2012 to August 2016. It is published to the DACCIWA database (http://baobab.sedoo.fr, last access: 10 June 2017) for the use of other modeling groups and for intercomparison with the gas flaring measurement flights of the EUFAR (European Facility for Airborne Research) mission APSOWA (Atmospheric Pollution from Shipping and Oil platforms of West Africa) project that was part of the DACCIWA aircraft campaign. APSOWA aims to characterize gaseous and particulate pollutants emitted by shipping and oil and gas extraction platforms off the coast of West Africa.

GFEI contains six inventories (GFEI-1 to GFEI-6) for the 364 flares within SWA. The results are presented in Figure 3.11 exemplarily for CO. The figures for the other species can be found in Appendix B.

GFEI-1 (Fig. 3.11a) uses as input for the flaring emission calculation the instantaneous (daily) data of source temperature and radiant heat. Furthermore, only the flares that are observed by VIIRS are considered (not the flares that are cloud covered). GFEI-2 (Fig. 3.11b) uses the instantaneous (daily) data of radiant heat but a monthly mean of source temperature for the respective flare, since the source temperature is sensitive to atmospheric conditions and therefore an average seems to be more robust. As GFEI-1 also this inventory only considers the flares that are directly observed by VIIRS. GFEI-3 (Fig. 3.11c) is derived by using the monthly mean data of source temperature and radiant heat. In this case all 364 flares are active at once using the flare specific monthly mean of source temperature and radiant heat. GFEI-4 (Fig. 3.11d) is equal to GFEI-2 in respect to the directly observed flares but adds for the cloud-covered flares (not observed but assumed to be active) the flare specific yearly emission average from GFEI-2. GFEI-5 (Fig. 3.11e) is the same as GFEI-4 but adds for the cloud-covered flares the flare specific DJF emission average from GFEI-2 instead of the yearly

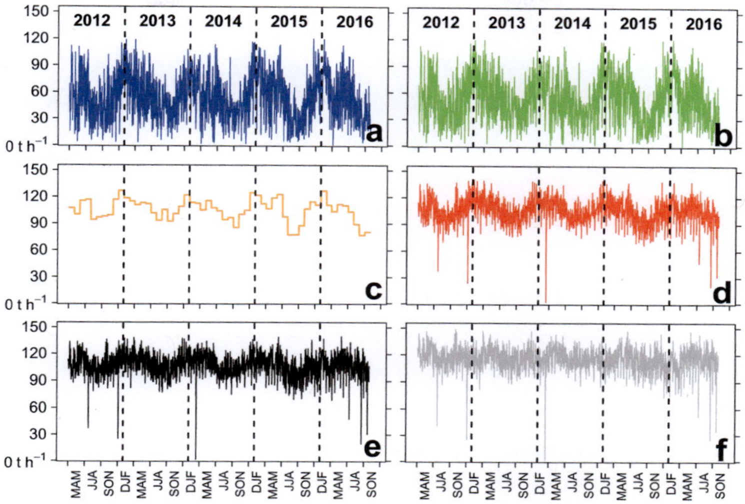

Figure 3.11: Gas flaring emission inventory (GFEI) 1-6 for CO (t h^{-1}), spatially integrated over SWA and capturing the time period March 2012 to August 2016. (a) GFEI-1: Using instantaneous (daily) data of source temperature and radiant heat by considering only the directly observed flares, (b) GFEI-2: Using instantaneous (daily) data of radiant heat but monthly mean of source temperature for the respective flare by considering only the directly observed flares, (c) GFEI-3: All 364 flares are active at once using their monthly mean source temperature and radiant heat, (d) GFEI-4: Sum of GFEI-2 and the emissions from cloud-covered flares (not observed but assumed to be active) using the flare specific yearly emission average from GFEI-2, (e) GFEI-5: As GFEI-4 but using the flare specific DJF emission average from GFEI-2 for the cloud-covered flares and (f) GFEI-6: As GFEI-5 but with removal of the annual cycle. The abscissa denotes the seasons and the vertical dashed lines separated the single years.

average. In the winter months, less atmospheric disturbances occur which lead to more reliable VIIRS observations. GFEI-4 and GFEI-5 consist of many assumptions but they consider the flaring variability from one day to another and the emissions from the cloud-covered flares in addition to avoid emission underestimations. Therefore it can be seen as a combination of instantaneous and "climatological" flaring data.

GFEI-6 (Fig. 3.11f) is based on GFEI-5 but cleans the time series from the annual cycle. Even by considering the cloud-covered flares as active, an annual cycle in the flaring emissions is visible with a maximum in the winter months and a minimum in the summer months. It is questionable whether this annual cycle is related to the atmospheric conditions that might disturb the remote sensing observations (e.g. via the WAM) or to reduced flaring activity in summer. However, the Nigerian oil production shows no distinct annual cycle (not shown). Figure 3.12 shows the SWA flares that are (a) observed (active), (b) not observed and not masked by clouds (inactive), (c) not observed and masked by clouds (assumed to be active) and (d) flares that are active or assumed to be active (a+c). Regarding Figure 3.12d, some years suppose to have an annual cycle indicating less flaring in summer (e.g. 2012 and 2013 but not 2015). So one reason for the strong annual cycle of the green curve is that the flaring activity in summer is reduced.

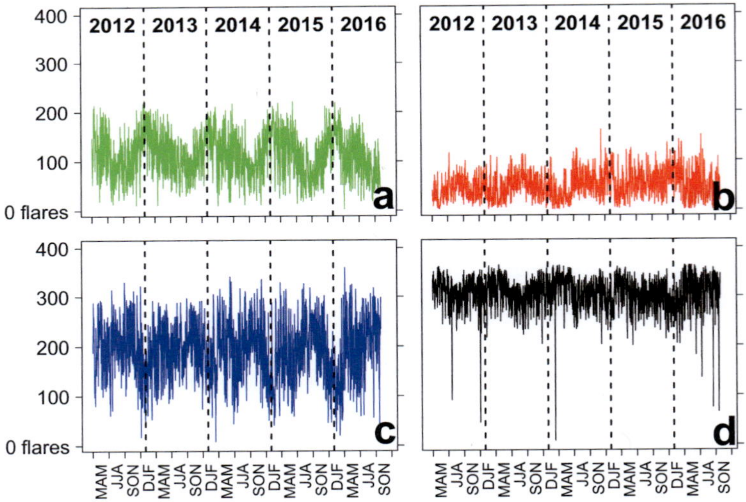

Figure 3.12: Number of flares within GFEI with the status (a) observed (active), (b) not observed and not masked by clouds (inactive), (c) not observed and masked by clouds (assumed to be active) and (d) active or assumed to be active (a+c). The abscissa denotes the seasons and the vertical dashed lines separated the single years.

It is also assumed that the atmospheric conditions contribute to this annual cycle. Within GFEI-6 (Fig. 3.11f) the annual cycle footprint was removed by adding for every month from March to November the monthly averaged emission difference with respect to DJF.

Every inventory includes the location, source temperature, radiant heat, flow rate and emission of CO, CO_2, SO_2, NO and NO_2. The application of the different inventories of GFEI depends on the research question. For a detailed comparison with in-situ observations, GFEI-2 is appropriate. For the use as point source emissions in an atmospheric model, GFEI-4, 5 or 6 can be used if daily files can be read from the model. GFEI-3 simplifies the process since only one file per month has to be read. For this study, the flare specific temporal mean (June-July 2016) of GFEI-4 is used in COSMO-ART.

The gas flaring emission inventory with its long temporal range and high spatiotemporal resolution, together with the detailed uncertainty assessment of the emission derivation methodology as described in Deetz and Vogel (2017), is a unique dataset and can support the understanding of the contribution of gas flaring to air pollution and its implications. Refer to Appendix A for the code and data availability.

3.1.4 Aerosol Feedback Mechanisms in COSMO-ART

COSMO-ART allows for the consideration of the ADE and AIE to assess the feedbacks between aerosol and the atmospheric state, in particular radiation, clouds and precipitation but also the atmospheric dynamics. In the following, the representation of ADE and AIE in COSMO-ART is described, including the method that is applied in this study to alter the aerosol amount for the conduction of aerosol sensitivity experiments.

Implementation of ADE

In COSMO the vertical profiles of short and longwave radiative fluxes were calculated for eight spectral bands (0.25-104.5 μm) using GRAALS (Ritter and Geleyn, 1992). The fluxes depend on temperature, pressure and the concentration of water, carbon dioxide, ozone and the size distributions of the aerosol particles. Based on the Mie theory the three optical properties extinction coefficient b, single scattering albedo ω and the asymmetry factor g were calculated to consider the ADE in COSMO-ART. These parameters depend on the aerosol size distribution and chemical composition (in particular their soot and water content). Since it is computational too expensive to calculate these properties at all grid points and every time step for all spectral bands, Vogel et al. (2009) implemented a parameterization, combining offline and online calculations to derive the three parameters. The offline part includes detailed Mie calculations for aerosol particle size distributions that were simulated with COSMO-ART without considering the ADE. From these results, the dependency of ω on the total wet aerosol mass m_l (μg m^{-3}) of the mode l and the dependency of g on the aerosol mass fraction f_l of the mode l (related to total aerosol mass including water) were determined. Via fitting techniques, the parameters $\tilde{b}_{k,l}$, $\tilde{\omega}_{k,l}$ and $\tilde{g}_{k,l}$ were derived for b_k, ω_k and g_k for all spectral bands k and the submicron modes l and were stored as a look-up table that is taken as representative and is used for all simulations considering the ADE. The look-up table is given in Vogel et al. (2009). This leads to the parameterization of the three optical properties: extinction coefficient

$$b_k = \sum_l 10^{-6}\, \tilde{b}_{k,l}\, m_l, \tag{3.36}$$

single scattering albedo in longwave range (k=4,5,6,7,8)

$$\omega_k = \sum_l \tilde{\omega}_{k,l}\, f_l \tag{3.37}$$

and shortwave range (k=1,2,3)

$$\omega_k = \tilde{\omega}_{k,AIT}\, f_{AIT} + \tilde{\omega}_{k,ACC}\, f_{ACC} + \tilde{\omega}_{k,SOOT}\, f_{SOOT}$$
$$+(2.6278\, f_{SOOT(AIT)} + 1)^{-1.8048}\, f_{AITS}$$
$$+(2.0611\, f_{SOOT(ACC)} + 1)^{-1.4309}\, f_{ACCS} \qquad (3.38)$$

and the asymmetry factor

$$g_k = \sum_l \tilde{g}_{k,l}\, f_l, \qquad (3.39)$$

where $f_{SOOT(AIT)}$ and $f_{SOOT(ACC)}$ are the soot fractions in the mixed Aitken and accumulation mode, respectively. The parameters with a tilde denote the offline part of the parameterization and the parameters without a tilde the online part considering the characteristics of the prognostic aerosol. The mode CANTHRO was neglected since it does not contribute remarkably to the extinction.
To assess the sensitivity of the ADE on the meteorological conditions, a namelist parameter was added in COSMO-ART within this study that allows to scale the total aerosol mass density within the ADE calculation by a factor F_{ADE}, affecting m_l in Equation 3.36. In this case the underlying aerosol distributions are preserved but the total mass is changed. All aerosol modes are changed uniformly by the factor. The changes of the aerosol amount are limited to the calculation of the ADE.

Implementation of AIE

The formation of cloud droplets is described in COSMO-ART via the Köhler theory (Köhler, 1936) regarding soluble aerosol particles and via the Frenkel-Halsey-Hill (FHH) adsorption theory (Sorjamaa and Laaksonen, 2007) for insoluble but wettable aerosol particles. The Koehler theory describes the equilibrium between water vapor and an aqueous solution droplet. The saturation s of air with respect to the liquid phase of water is the ratio of vapor pressure e_w to saturation vapor pressure e_w^0 and the supersaturation s_s is defined as

$$s_s = s - 1 = \frac{e_w}{e_w^0} - 1. \tag{3.40}$$

Two counteracting effects determine the vapor pressure over a droplet: the Kelvin effect, that considers the vapor pressure increase with increasing curvature of the droplet, and the Raoult effect describing the vapor pressure decrease with increasing concentration of solute mass (e.g. sodium chloride). The superposition of both effects leads to an equilibrium supersaturation $s_{s,e}$ described by the Köhler equation

$$s_{s,e} = \frac{A}{d_D} - \frac{B}{d_D^3}, \tag{3.41}$$

with

$$A = \frac{4\,M_w\,\sigma_w}{R\,T\,\rho_w} \tag{3.42}$$

and

$$B = \frac{6\,n_{sol}\,M_w}{\pi\,\rho_w}, \tag{3.43}$$

where d_D is the diameter of the aqueous solution droplet, M_w the water molecular mass, σ_w the water surface tension, ρ_w the water density, R the universal gas constant, T the temperature and n_{sol} the amount of substance (mol) of solute. In Equation 3.41 the term A/d_D denotes the Kelvin effect (curvature term) and B/d_D^3 the Raoult effect (solute term). Figure 3.13 illustrates the Köhler curve for a sodium chloride particle with a diameter of 0.05 μm (solid red) as a superposition of the Kelvin effect (blue) and Raoult effect (green). The smaller the droplet the greater the curvature and therefore the greater the vapor pressure (blue curve) and the smaller the droplet the higher the concentration of solute in the droplet and therefore the lower the vapor pressure (green curve). With increasing droplet diameter, the curves converge asymptotically to $s_s = 0$ because the curvature and the concentration

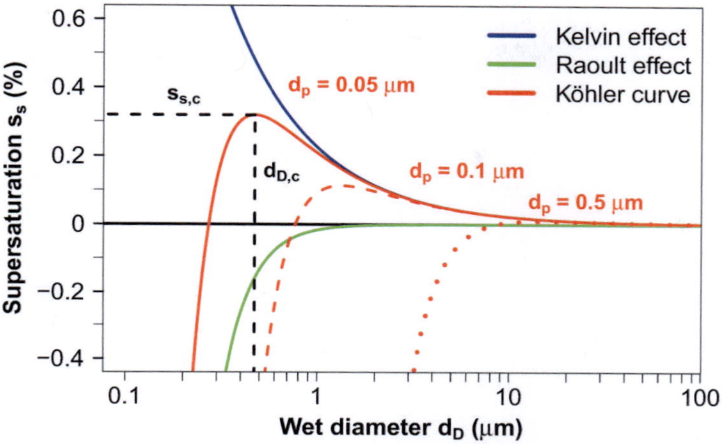

Figure 3.13: Köhler curve (equilibrium supersaturation $s_{s,e}$ for a spherical sodium chloride particle with a dry diameter d_p of 0.05 μm (solid red), 0.1 μm (dashed red) and 0.5 μm (dotted red) at 293 K. For the solid red Köhler curve, the contribution from the Kelvin effect (blue) and Raoult effect (green) as well as the location of the critical diameter $d_{D,c}$ and critical supersaturation $s_{s,c}$ (dashed black) are shown. The parameters A and B are approximated according to Seinfeld and Pandis (2006a).

of solute mass decrease. Both effects increase with decreasing droplet diameter, but the Raoult effect much faster ($\sim 1/d_D^3$). The blue curve in Figure 3.13 clearly illustrates that very small cloud droplets consisting of pure water (without solute) cannot be formed in the atmosphere since this would require very high supersaturations. A CCN is necessary to reduce the vapor pressure. The Köhler curve (red solid curve) shows a maximum at the critical diameter $d_{D,c}$ and the critical supersaturation $s_{s,c}$ which can be calculated via the first derivate of Equation 3.41

$$d_{D,c} = \sqrt{\frac{3B}{A}} \tag{3.44}$$

and

$$s_{s,c} = \sqrt{\frac{4A^3}{27B}}. \tag{3.45}$$

Droplets below $d_{D,c}$ are in a stable state and will return to the equilibrium state after small changes of their diameter. Droplets that exceed $d_{D,c}$ will further grow because lower supersaturations are necessary (see Fig. 3.13). Infinite growth does not occur since the condensational growth is linked to a reduction of the water vapor content in the atmosphere and therefore a reduction of the supersaturation. The maximum supersaturation $s_{s,max}$ in an ascending air parcel is the supersaturation that is available for the condensational growth. Only if $s_{s,c} \leq s_{s,max}$ the droplets can grow further. Since $s_{s,c}$ is a function of the mass and chemical composition of the aerosol particle, the particle is labeled *activated* as soon as it exceeds $d_{D,c}$. The activation transforms the CCN into cloud droplets.

The prognostic aerosol of COSMO-ART, including the information about the aerosol size distribution and the aerosol chemical composition, is linked to the activation concept described above via the parameterization of Fountoukis and Nenes (2005). Within Fountoukis and Nenes (2005) the CCN concentration as a function of s_s is calculated, assuming ideal solutions of the aerosol particles. Subsequently, $s_{s,max}$ is derived via Equation 3.46 (Seinfeld and Pandis, 2006a) by including the CCN in an adiabatic parcel with a constant updraft velocity w and considering the equilibrium between the water vapor increase from the adiabatic cooling (first term on right hand sight of Eq. 3.46) and the water vapor decrease from the condensational growth (second term right hand side of Eq. 3.46)

$$\frac{ds}{dt} = \alpha w - \gamma \frac{dW}{dt}, \tag{3.46}$$

with

$$\alpha = \frac{g\, M_w\, \Delta H_v}{c_p\, R\, T^2} - \frac{g\, M_a}{R\, T} \tag{3.47}$$

and

$$\gamma = \frac{P\, M_a}{e_w^0\, M_w} - \frac{M_w\, \Delta H_v^2}{c_p\, R\, T^2}, \tag{3.48}$$

where ΔH_v is the latent heat of condensation, T is the parcel temperature, g is the gravitational acceleration, M_a is the molecular weight of air, c_p is the specific heat capacity of air, P is the ambient pressure and dW/dt the rate of condensation onto the droplets.

The CCN grow instantaneously to $d_{D,c}$ when $s_{s,c}$ is reached. To avoid an underestimation of the cloud droplet number concentration (CDNC) N_c, the decelerated growth of giant CCN (kinetic limitation) is considered by using Barahona et al. (2010). $s_{s,max}$ is reached when the two terms on right hand side of Equation 3.46 are balancing each other. For the insoluble but wettable aerosol particles, the parameterization of Kumar et al. (2009) is used that follows the FHH adsorption theory. The equilibrium of the insoluble particle with the ambient water vapor is given as

$$s_{s,e,FHH} = \alpha_w \, exp\left(\frac{4\,\sigma_i\,M_w}{R\,T\,\rho_w\,d_p}\right) - 1, \tag{3.49}$$

with

$$\alpha_w = exp\left(-A_{FHH}\Theta'^{-B_{FHH}}\right), \tag{3.50}$$

where α_w is the activity of the water in the particle, σ_i is the surface tension at the interface between particle and gas, Θ' the number of adsorbed water molecules divided by the number of molecules in a water layer, A_{FHH} and B_{FHH} are compound specific empirical constants. In COSMO-ART the values $A_{FHH} = 2.5$ and $B_{FHH} = 1.2$ are used for mineral dust. COSMO-ART only treats mineral dust via Kumar et al. (2009). Soot is not treated via the FHH theory since the number of soot particles in the pure soot mode SOOT is small and these particles are quickly transferred to the mixed submicron modes during the aging process that provides the soot particles with a soluble coat. Therefore soot is treated via the Köhler theory.

Bangert (2012) coupled the prognostic aerosol of COSMO-ART and the activation mechanism with the TMMS of Seifert and Beheng (2006). This coupling includes the use of parameterized subgrid scale vertical velocities since COSMO-ART, as a regional model, is not able to resolve the small scale updraft velocities that drive the activation of CCN. To use only the grid scale velocity w would lead to an underestimation of the activated CCN and therefore an underestimation of the CDNC. According to Morales and Nenes (2010), the subgrid scale vertical velocities w' are described with a Gaussian probability distribution function

$$P_w(w') = \frac{1}{\sqrt{2\pi}\sigma_w} exp\left(-\frac{(w'-w)^2}{2\sigma_W^2}\right). \tag{3.51}$$

The mean of $P_w(w')$ is set to w and the standard deviation σ_w is derived from the square root of the turbulent kinetic energy. By using w' and Equation 3.51, a weighted average of the activated particles is calculated. The activated particles (nucleated CDNC $N_c^*(s_{s,max})$) from the FHH or Köhler theory are transformed to a nucleation rate J_{act} which is then used in the TMMS. For the three cases (a) new cloud (or $N_c^*(s_{s,max}) < 10\,cm^{-3}$), (b) in an existing cloud and (c) at the cloud base, the following equations are used for the transformation:

$$J_{act} = \begin{cases} max\left(\frac{N_c^*(s_{s,max})}{\Delta t}, 0\right) & new\ cloud \\ max\left(\frac{N_c^*(s_{s,max}) - N_c(t-1)}{\Delta t}, 0\right) & in\ existing\ cloud \\ -\frac{\partial}{\partial z} N_c^*(s_{s,max})\, w + \frac{\partial}{\partial z} K \frac{\partial}{\partial z} N_c^*(s_{s,max}) & at\ cloud\ base, \end{cases} \tag{3.52}$$

where Δt is the model timestep of COSMO-ART, $N_c(t-1)$ the CDNC of the previous timestep and K the turbulent diffusion coefficient. The pre-existing cloud droplets $N_c(t-1)$ for the activation in an existing cloud are taken as giant CCN linked with a decelerated growth. The activation rate at the cloud base is a function of the vertical advection and diffusion of particles into the cloud.

Although this study focus on warm clouds and not on mixed-phase or cold clouds, also the ice nucleation is considered in the model to avoid misrepresenting processes which might have an influence (e.g. the passage of MCS). The ice nucleation is parametrized by Philipps et al. (2008). The formation of cirrus and the competition between homogeneous and heterogeneous freezing is described according to Barahona and Nenes (2009a,b). To assess the sensitivity of the AIE on the meteorological conditions, analogously to F_{ADE} regarding ADE, a namelist parameter F_{AIE} was added in COSMO-ART within this study that allows to scale the aerosol number concentration. This factor is independent from F_{ADE}. Also for F_{AIE} the underlying aerosol distributions are preserved but the total number is changed. All aerosol modes are changed uniformly by the factor. The changes of the aerosol number are limited to the aerosol activation mechanism. Figure 3.14 illustrates exemplarily the change of the aerosol number size distribution by two values of F_{AIE}.

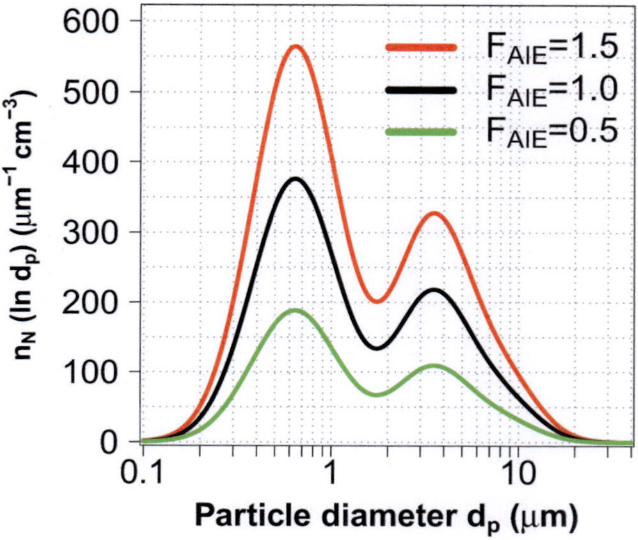

Figure 3.14: Example of an aerosol number size distribution (μm^{-1} cm^{-3}) altered by F_{AIE} within the aerosol activation mechanism to assess the impact of changing aerosol amounts on AIE.

3.1.5 Simulation Setup

Three domains are used for this study (Fig. 3.15) that are denoted as Domain 1 (blue, D1 hereafter), Domain 2 (green, D2 hereafter) and Domain 3 (red, D3 hereafter). The characteristics of the three domains are described in Table 3.7. D1 is applied for the operational aerosol/chemistry forcasts from 8 May to 31 July 2016. The simulation period in May 2016 was used for testing the forecast system and model spin-up. In the two month June-July 2016 the forecasts supported the DACCIWA measurement campaign. For D1 the OMMS was used. In this case the AIE is constrained by a spatiotemporally constant prescribed CDNC of N_c=500 cm^{-3} and ADE is considered by using the aerosol climatology of Tegen et al. (1997).

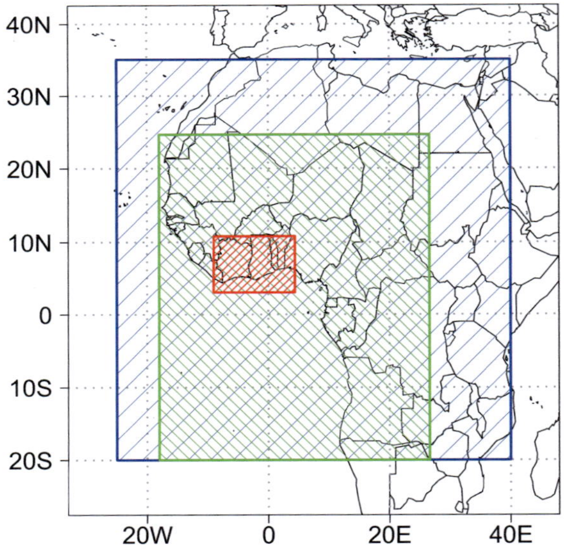

Figure 3.15: COSMO-ART simulation domains applied for this study. Blue, green and red refer to Domain 1 (D1), Domain 2 (D2) and Domain 3 (D3), respectively. D1 and D2 are nested in global ICON forecasts, whereas D3 is nested in D2.

Table 3.7: Overview of the COSMO-ART simulation setup for D1, D2 and D3.

Characteristics	**D1**	**D2**	**D3**
Time period	8 May - 31 July 2016	25 June - 3 July 2016	2 July - 3 July 2016
Simulation domain	25°W-40°E, 20°S-35°N	18.0°W-26.6°E, 20.0°S-24.6°N	9.0°W-4.4°E, 3.0°N-10.8°N
Grid mesh size (km)	28	5	2.5
Number of vertical levels	50	50	80
Met. boundary conditions	ICON	ICON	D2
Convection parameterization	Yes	No	No
Cloud microphysics	OMMS	TMMS	TMMS
ADE	Climatology	Climatology	Prognostic aerosol
AIE	Prescribed N_c	Prescribed CCN	Prognostic aerosol
Aerosol/Chemistry boundary conditions	MOZART	MOZART	D2

The realization and outcomes of the D1 forecasts are described in Section 5. In contrast to D1, D2 and D3 are applied for conducting a detailed case study, analyzing ADE and AIE on 2 and 3 July 2016. D2 is the intermediate step as a basic for realizing D3. Therefore D2 results are only sporadically shown in this study. The main focus is on D3.

For D2, the convection parameterization was switched off. Although, we are aware of the recommendation that model realizations with explicit convection should be connected with grid mesh sizes below 3 km, the explicit convection was used with a 5 km grid mesh size that significantly improved the model performance (Sect. 6.1.1). Pearson et al. (2013) showed that even with a grid

mesh size of 12 km, explicit convection leads to a better model representation of the diurnal cycle of tropical convection compared to the results using the convection parameterization (see also Marsham et al., 2013). D2 includes the use of the TMMS but without prognostic aerosol due to computational cost. Regarding the AIE the climatology of Segal and Khain (2006) for the continental case was used that prescribes a spatiotemporally constant aerosol number concentration of 1700 cm^{-3} in the lowest 2000 m and an exponential concentration decrease above. The ADE is considered as for D1.
In D3 the TMMS was combined with the prognostic aerosol of COSMO-ART. Therefore the aerosol is variable in space and time and consistently considered in ADE and AIE. However, it has to be considered that the CDNC in D2, resulting from the prescribed continental aerosol, can be advected into D3 via the nesting and can lead to inconsistencies. On the other hand the predominant near-surface atmospheric flow (monsoon flow) coming from D2 into D3, originates from the Gulf of Guinea, and the coastal region including the cities as major aerosol sources are located within D3. Therefore it is assumed that these inconsistencies are negligible. The D3 simulations are conducted by a variation of the aerosol in ADE and AIE via the factors F_{ADE} and F_{AIE}, introduced in Section 3.1.4. The selection of the factors and the outcome of the realizations are presented in detail in Section 6. For all simulations, ICON operational forecasts of the German Weather Service (DWD) were used with a grid mesh size of R03B07 (about 13 km). Additionally, MOZART forecasts (MOZART, 2017a) and data from the MOZART archive (MOZART, 2017b), with a grid mesh size of 280x213 km, were used as boundary data for the atmospheric pollutants in D1 and D2/D3, respectively.

4 Data

For this study, a number of datasets was used on the one hand as input for COSMO-ART and on the other hand for evaluating the COSMO-ART performance. The latter includes remote sensing data and data that is obtained during the DACCIWA measurement campaign. These datasets are listed and introduced in the following to provide an overview, to facilitate the comprehension of the subsequent data application and to allow for reproducibility of the outcomes.

4.1 COSMO-ART Boundary and Initial Data

Aside from the boundary data of meteorology and atmospheric composition, the data applied for the COSMO-ART simulation are primarily related to the different emissions of pollutants, described in detail in Section 3.1.3. Therefore this part confines to the listing of the used datasets in Table 4.1.

Table 4.1: Overview of the datasets applied for the COSMO-ART realizations. The type of dataset is categorized in ground based (G), model (M) and space-borne (S). The grid mesh size (°) refers to the resolution of the original datasets. The gas flaring dataset includes the remote sensing observations of VIIRS (2016) and is therefore assigned to space-borne and the spatial resolution of VIIRS.

Data source	Type	Parameter	Grid mesh size (°)
Lana et al. (2011)	G	Monthly mean DMS fluxes	1x1
HWSD (2012)	G	USDA soil types	0.0083x0.0083
EDGAR (2010)	G	Pollutant emissions	0.1x0.1
ICON operational forecasts	M	Meteorological boundary	0.12 (R03B07)
MOZART (2017a) (forecasts)	M	Atmospheric composition boundary (D1)	2.5x1.9
MOZART (2017b) (archive)	M	Atmospheric composition boundary (D2, D3)	2.5x1.9
CCSM (2015)	S	Plant functional types (PFTs)	0.05
Deetz and Vogel (2017)	S	Gas flaring pollutant emissions	0.007
GlobCover (2009)	S	Landuse classes	0.002
CAMS (2017a)	S	GFAS biomass burning pollutant emissions	0.1x0.1

4.2 DACCIWA Campaign Related Data for Model Evaluation

Within the DACCIWA project, an extensive field campaign took place in June-July 2016 at SWA to set about the problem of observational data shortage that impedes the understanding of the meteorological processes and mechanisms in the research area (Knippertz et al., 2015a). The time period June-July was selected to capture the onset of the WAM and a period characterized by increased cloudiness. Originally, the campaign was planned for 2015 but had to be postponed due to the Ebola epidemic in that year. June-July 2016 denotes the general frame of the DACCIWA field campaign. The components of the field campaign capture subsets of this time period (Flamant et al., 2017) as denoted in Table 4.2.

Table 4.2: The DACCIWA measurement campaign and their sub-periods. The intensive operation period (IOP) 6 is capturing the time period that is used for the aerosol feedback case study with COSMO-ART.

Measurement campaign	Time period
DACCIWA measurement campaign	01 June - 31 July 2016
DACCIWA ground-based measurement campaign	13 June - 31 July 2016
DACCIWA ground-based measurement campaign, IOP 6	02 July - 03 July 2016
DACCIWA aircraft special observing period (ASOP)	27 June - 17 July 2016

The DACCIWA ground-based measurement campaign, as the first component, encompassed the time period from 13 June to 31 July 2016, including the three supersites Kumasi (Ghana), Savè (Benin) and Ile-Ife (Nigeria) (Fig. 2.1). The ground-based observations consider measurements of the near-surface meteorological parameters and the profiling of the meteorological characteristics in the PBL and the entire troposphere using ground-based instruments and radiosoundings. 15 intensive operation periods (IOPs) were defined that are related to a launching of radiosondes in higher temporal frequencies (1.5-3 h). IOP 6, denoted to the time period 2-3 July 2016 (Tab. 4.2), coincide with the aerosol feedback case study conducted in this work. Table 4.3 summarizes the ground-based data that are applied for an evaluation of COSMO-ART. A complete overview of the DACCIWA ground-based measurement campaign, their supersites, instrumentation and a first insight into the available data is presented in Kalthoff et al. (2017).

To assess the perfomance of COSMO-ART, measurements of basic meteorological parameters (e.g. near-surface temperature and humidity) are used. Furthermore, the ceilometer observations provide highly valuable insights in the characteristics of the cloud base height and their temporal evolution. The profiles of wind speed and relative humidity from doppler lidar and radiosondes, respectively, have been used to evaluate the specification of the NLLJ and NLLS. For the quantification of the amount of atmospheric pollutants, the GRIMM aerosol spectrometer, measuring the aerosol mass in size bins between 0.23 and 32 μm, a sun photometer, providing the total AOD, and data from the chemistry tower, providing among other the volume mixing ratio of CO and O_3, were used. Unfortunately, the GRIMM device failed after a few days of operation. Further information on the PBL aerosol is provided by the *Unmanned Aerial System Carolo P360 Application of Light-weight Aircraft for Detecting in-situ Aerosol* (UAS Carolo P360 ALADINA, Altstädter et al., 2015). ALADINA measured vertical profiles of black carbon (BC) mass concentrations that are compared with results of the COSMO-ART realizations.

The second component of the DACCIWA measurement campaign is the DACCIWA aircraft special observing period (ASOP) from 27 June to 17 July 2016 (Flamant et al., 2017). Within the ASOP, three research aircrafts were used to capture the characteristics of aerosol, chemistry and cloud properties: the DLR Falcon 20 (Germany), the SAFIRE ATR 42 (France) and the British Antarctic Survey (BAS) Twin Otter (UK).

Table 4.3: Overview of the ground-based observational data that is dedicated to the DACCIWA measurement campaign and that is used for the evaluation of COSMO-ART.

Data source	Parameter	Instrument
Savè supersite	Near-surface temperature	Energy balance station
Savè supersite	Near-surface RH	Energy balance station
Savè supersite	Surface shortwave radiation	Energy balance station
Savè supersite	Precipitation	Energy balance station
Savè supersite	Near-surface wind speed	Energy balance station
Savè supersite	Vertical profile of wind speed	Doppler lidar
Savè supersite	Cloud base height	Ceilometer
Savè supersite	Vertical profile of RH	Radiosondes
Savè supersite	Aerosol mass concentration	GRIMM aerosol spectrometer
Savè supersite	CO volume mixing ratio	Chemistry tower
Savè supersite	O_3 volume mixing ratio	Chemistry tower
AERONET (2017) (Savè)	Total AOD	Sun photometer
UAS ALADINA	BC mass concentration	Aethalometer

An overview of the ASOP, the aircraft payload and first outcomes are published in Flamant et al. (2017). For this study, primarily the trace gas concentration measurements of the Twin Otter are used for a comparison with COSMO-ART. In Table 4.4 the used aircraft data is summarized. The flight planning was supported by tailored model forecasts of the atmospheric state as well as the distribution of atmospheric pollutants (Knippertz et al., 2015a). In this regard and as a part of this study, operational aerosol/chemistry forecasts for SWA were conducted and provided. The realization and outcomes of these forecasts are presented in Section 5, together with the evaluation against the observations introduced in this data section.

Table 4.4: Overview of the airborne observational data that is dedicated to the DACCIWA measurement campaign and that is used for the evaluation of COSMO-ART.

Data source	Parameter	Instrument
Twin Otter	Temperature	Rosemount non-deiced temperature sensor
Twin Otter	Pressure	Static pressure from the aircraft probe
Twin Otter	CO	Aerolaser AL5002
Twin Otter	NO	Air Quality Design fast NOx
Twin Otter	NO_2	Air Quality Design fast NOx
Twin Otter	NO_x	Air Quality Design fast NOx
Twin Otter	O_3	2B Technologies 205
Twin Otter	SO_2	TEI 43i
Twin Otter	CDNC	CDP
Twin Otter	Cloud droplet effective radius	CDP
Twin Otter	Cloud liquid water concentration	CDP

Regarding the ground-based observations, this study focuses on Savè with its sophisticated measurement system combining the KITcube (Kalthoff et al., 2013) and trace gas observations from Université Toulouse III - Paul Sabatier, Laboratoire d'Aérologie (UPS). Also trace gas observations and properties of liquid clouds observed from the *British Antarctic Survey (BAS) Twin Otter* aircraft are used for intercomparison. The CDP (Cloud droplet probe) is a wing mounted canister instrument including a forward-scatter optical system to measure the cloud droplet spectrum between 2-50 μm with a frequency of

1 Hz. The aerosol optical depth (AOD) is assessed by using ground based data (AERONET, 2017), space-borne data of MODIS (Hsu et al., 2013) and model data of CAMS (2017b).

4.3 Further Data for Model Evaluation

In addition to the measurement data obtained within the DACCIWA measurement campaign, further datasets are used for evaluating the COSMO-ART results (Tab. 4.5). This includes the AOD observations at the Aerosol Robotic Network (AERONET, 2017) stations Ilorin in southwestern Nigeria and Koforidua in southeastern Ghana, complementing the observations at Savè. By using model data from Copernicus Atmosphere Monitoring Service (CAMS, 2017b), an AOD analysis, separated in the components mineral dust, sea salt and anthropogenic aerosol, is possible. The AOD evaluation is completed by the use of remote sensing observations from MODIS (2017). To assess the modeled cloud cover, data from MSG/SEVIRI (Meteosat Second Generation/Spinning Enhanced Visible and Infrared Imager, CMSAF, 2017) is applied. Although this product has a high temporal resolution (geostationary), it is known that the remote sensing retrievals have serious problems in detecting low-level clouds above land at night due to the small temperature contrast between land surface and cloud top (van der Linden et al., 2015), which has to be considered in the interpretation of the results. The modeled precipitation is compared to the Tropical Rainfall Measuring Mission Multi-Satellite Precipitation Analysis, the Near Real-Time processing stream product (TRMM (TMPA-RT), Huffman, 2016).

Table 4.5: Overview of the observational data that is not dedicated to the DACCIWA measurement campaign and that are used for the evaluation of COSMO-ART. The type of observation is categorized in ground based (G), model (M) and spaceborne (S).

Data source	Type	Parameter	Instrument/Variable	Grid mesh size (°)
AERONET (2017) Ilorin	G	Total AOD	Sun photometer	-
AERONET (2017) Koforidua	G	Total AOD	Sun photometer	-
CAMS (2017b)	M	Mineral dust AOD Sea salt AOD Anthropogenic AOD	CAMS Near-real-time	0.71x0.71
MODIS (2017) Terra	S	Total AOD	MOD04_L2 Deep Blue Aerosol	0.09x0.09
MODIS (2017) Aqua	S	Total AOD	MYD04_L2 Deep Blue Aerosol	0.09x0.09
MSG/SEVIRI (CMSAF, 2017)	S	Cloud cover	Fractional cloud cover (CFC)	0.027
TRMM (TMPA-RT) (Huffman, 2016)	S	Precipitation	TRMM_3B42RT_7	0.25x0.25

5 Aerosol/Chemistry Forecasts for the DACCIWA Measurement Campaign

To support the decision-making of research aircraft flight tracks within the DACCIWA aircraft special observing period (ASOP) from 27 June to 17 July 2016, we have developed a COSMO-ART forecast system to provide aerosol/chemistry fields to the DACCIWA community. The forecast system contains the full chain of preprocessing the input data, conducting the forecast, visualizing the results and providing the visualizations to the user.

However, the forecast supply is only one of the objectives we pursue with the model realizations. Usually, only short simulation periods on the order of magnitude of days were realized with COSMO-ART. The availability of model data on the order of magnitude of months is exceptional, allowing to extend the focus to research questions that include the validation and model intercomparison in terms of aerosols, chemistry and meteorology as well as the derivation of aerosol/chemistry statistics to obtain further insight in the characteristics of SWA and to assign the case study model results (domain D2 and D3 in Fig. 3.15) to a greater scale (e.g. when discussing the origin of pollution plumes).

The forecasts can be named operational. Although the model is used in research mode without assimilation, the forecasts were conducted continuously for nearly three month following a strict timeline aiming on providing the products on schedule. The continuous forecasts were initiated at 8 May and were active till 31 July 2016. In this time period 85 forecasts with in total 4845 simulation hours were realized. The forecast period covers the ASOP as well as the entire DACCIWA campaign period in June-July 2016. This modeling effort marks the first operational application of COSMO-ART with full chemistry and aerosol dynamics.

In the following, the forecast processing is described in Section 5.1, Section 5.2 illustrates the distribution of the forecast products and a more detailed data analysis in terms of forecast statistics and validation is presented in Section 5.3. In June-July 2014 and 2015, dry runs of the daily aircraft planning were conducted by the DACCIWA principle investigators in preparation of the actual campaign. The author contributed in 2015 i.a. by analyzing and presenting results of atmospheric composition forecasts.

5.1 Forecast Realization (Domain D1)

The forecasts were realized at the Forschungshochleistungsrechner ForHLR (Phase) I of the Steinbuch Centre for Computing (SCC) at KIT. One forecast required in total 1890 CPU hours. By using 240 processors, the COSMO-ART forecast computing time is 8 hours. The simulation domain (D1) is shown in blue in Figure 3.15 and the setup is indicated in Table 3.7. The meteorological boundary data is provided from the DWD as 3 hourly global ICON forecasts (R03B07) for 57 forecast hours. As aerosol/chemistry boundary the MOZART-4/GEOS-5 global forecasts (MOZART, 2017b) as 6 hourly data for 48 forecast hours were used. The aerosol/chemistry initialization with MOZART is only done for the very first forecast. All subsequent simulations are initialized via the previous COSMO-ART forecast in terms of aerosols and chemistry. In addition, ECMWF MACC gave us access to the near-realtime biomass burning emissions from GFASv.1.2 with a temporal offset of 1 day.

The COSMO-ART simulation delivered a 3 hourly forecast on pressure levels for a 57 h lead time. The model input and output data (especially ICON, COSMO-ART and INT2LM-ART) are stored on the Large Scale Data Facility (LSDF) of SCC and can be made available on demand for research purposes.

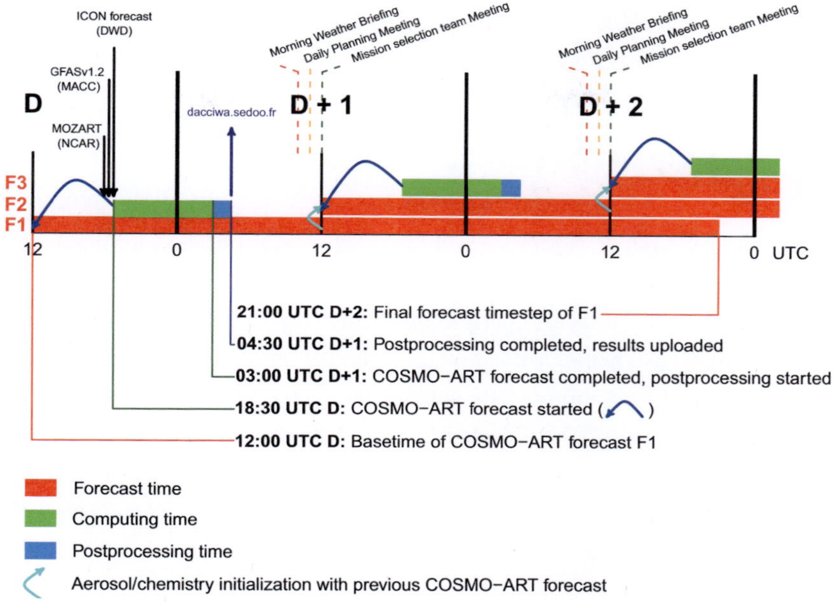

Figure 5.1: Scheme of the COSMO-ART forecast processing, exemplarily for the three forecast realizations F1, F2 and F3 (red bars), initialized at D, D+1 and D+2, respectively. One forecast consists of 57 forecast hours. The computing time is shown as green bar and the data postprocessing time (including raw data processing, plotting, uploading of the visualization to dacciwa.sedoo.fr and data storage) as blue bar. The light blue arrows indicate the aerosol/chemistry initialization via the previous COSMO-ART forecast (warm start). As updated external input, the meteorological forecasts of ICON (DWD), the GFAS fire emissions (MACC) and the aerosol/chemistry boundary fields of MOZART (NCAR) are used. The beginning of the *Morning Weather Briefing*, *Daily Planning Meeting* and the *Mission selection team meeting* at Lomé (location of aircraft coordination) is denoted by the dashed red, orange and green line, respectively.

Figure 5.1 shows the forecast processing schematically for two and a half days. The forecasts (red bars) were initialized every day starting at 12 UTC. After a computing time of 8 hours (green bars) the postprocessing takes place (blue bars). The results were uploaded to dacciwa.sedoo.fr, the public available campaign webserver. The beginning of the different daily DACCIWA meetings at Lomé, for which the forecasts are produced, are shown in red (10 LT:

Morning Weather Briefing), orange (11 LT: Daily Planning Meeting) and green (12 LT: Mission selection team Meeting).

5.2 Supply of the Forecast Results

The simulation results are processed via a set of R scripts to plot horizontal distributions as well as vertical transects of atmospheric pollutants for SWA. A 57 h COSMO-ART forecast finally delivers 3600 figures, including wind speed and direction, PBL height estimation, mass and number concentration of black carbon, mineral dust and sea salt, sulfate, submicron particles, ozone, carbon monoxide, NO_x, sulfur dioxide and isoprene. Additionally, to allow for a better overview, especially when the aircraft planning team is confronted with a poor internet connection in Lomé, a PDF was provided, collating a 116 page daily summary of the forecast.

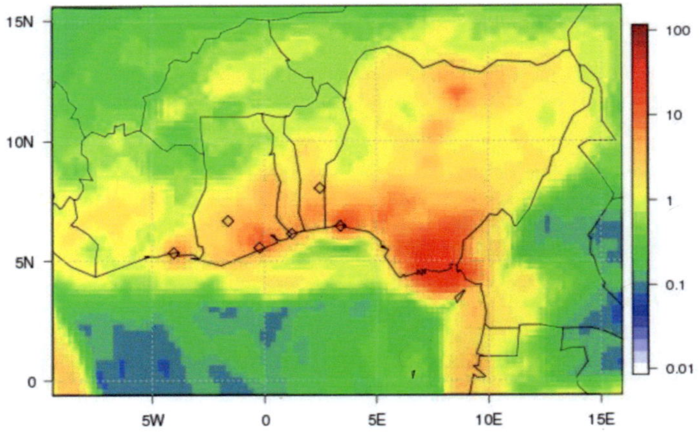

Figure 5.2: COSMO-ART forecast of NO_x volume mixing ratio at 1000 hPa (sum of NO and NO_2) (ppbv) on 3 July 2016 12 UTC (with a lead time of 48 h) for the SWA domain. Selected SWA city locations are indicated with black diamonds.

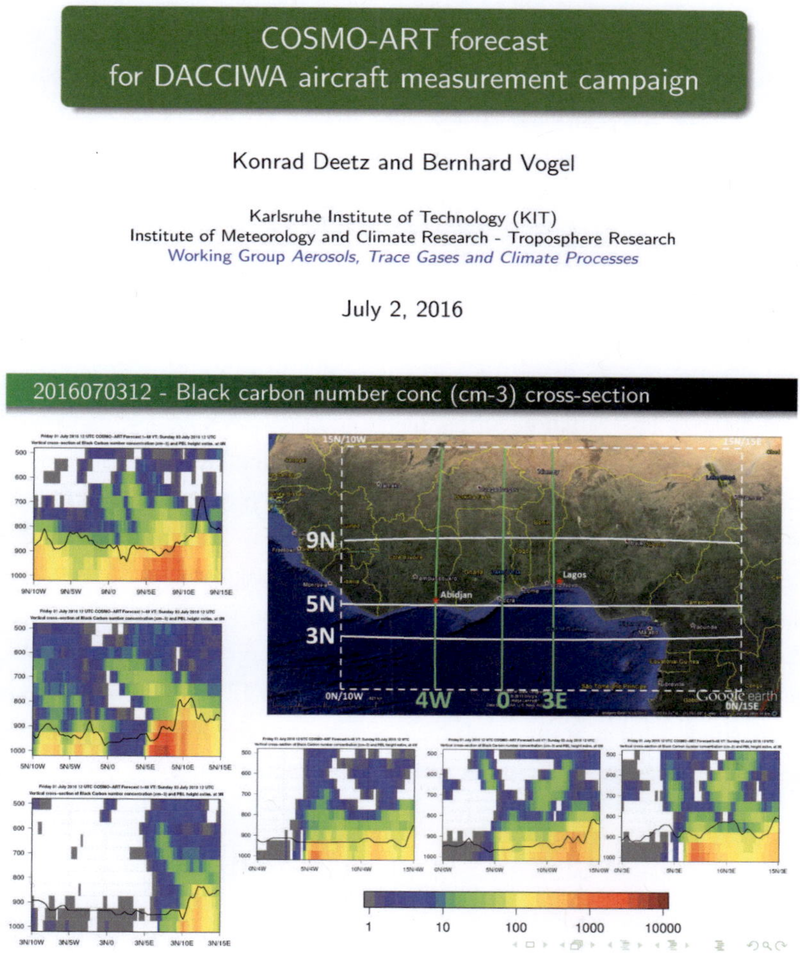

Figure 5.3: COSMO-ART forecast summary PDF, title page (top) and black carbon number concentration (cm^{-3}) (bottom) on 3 July 2016, 12 UTC (with a lead time of 48 h). Six vertical transects are given, three zonal (left column: 9°N, 5°N and 3°N from top to bottom) and three meridional (bottom row: 4°W, 0° and 3°E from left to right). The location of the cross sections is illustrated by the top right map. The black line within the vertical transects indicates the estimated COSMO PBL height.

The slides are in a format suitable for a beamer presentation to allow for direct application in the daily briefings. During the campaign, further vertical transect visualizations were added, customized according to the requests of the researchers on site. The figures and summary documents are uploaded to the dacciwa.sedoo.fr webserver (see http://dacciwa.sedoo.fr/source/indexItem.php?current=20160601&nav=COSMO-ART, last access: 9 September 2017). Exemplarily for 3 July 2016, two selected figures from the standard forecast output are presented. Figure 5.2 shows the NO_x volume mixing ratio (ppbv) at 1000 hPa on 3 July 2016 12 UTC. The plumes from the cities (black diamonds) were transported with the monsoon flow in northeasterly direction. This near-surface transport pattern shows less variation during the entire DACCIWA campaign. Over the Niger Delta (6-9°W and 4-7°N) very high concentrations are visible, due to anthropogenic emissions from Port Harcourt and oil production. Over the Gulf of Guinea, NO_x is related to shipping emissions as well as the transport from the southeasterly coast. Also biomass burning emissions from central Africa play a role. Figure 5.3 shows selected slides of the forecast summary PDF. The bottom figure presents the vertical transect overview regarding the black carbon number concentration, valid for 3 July 2016 12 UTC. The assignment of the single vertical transects to their geographical location is simplified by the top right map including the corresponding transect courses.

5.3 Forecast Statistics and Validation

In the following, observational data from the DACCIWA measurement campaign (Sect. 4.2) and additional data (Sect. 4.3) are used to evaluate the modeled meteorology and atmospheric composition in the framework of the COSMO-ART forecasts (D1), locally at Savè (Sect. 5.3.1 and 5.3.3, respectively) and on a greater spatial scale for SWA (Sect. 5.3.2 and 5.3.4, respectively).

5.3.1 Meteorological Quantities Locally at Savè Supersite

As an introductory overview, Figure 5.4 shows the NLLS over Savè as a diurnal cycle from 18 UTC to 18 UTC between 14 June and 30 July 2016. Figure 5.4b is the observed 30-min cloud-base fraction (%) for the lowest 1000 m AGL derived from ceilometer data (Kalthoff et al., 2017, adapted). Additionally, Figure 5.4a shows the COSMO-ART vertically integrated cloud water (g m^{-2}) over the lowest 1000 m AGL. The black lines separate the meteorological phases indicated by Knippertz et al. (2017) (Tab. 2.2). The hatched area is identified as the monsoon *Onset phase* with the *Pre-onset phase* before 21 June and the *Post-onset phase* after 21 June (Knippertz et al., 2017).

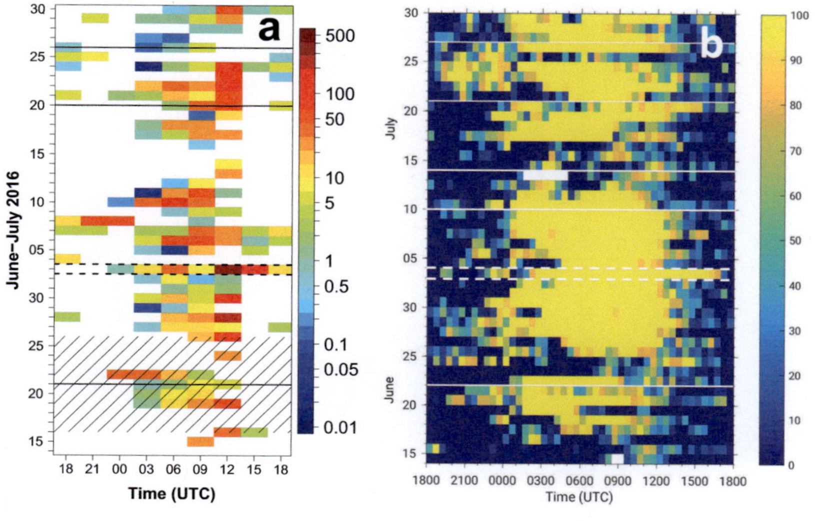

Figure 5.4: NLLS timeseries at Savè supersite, covering the time period 14 June to 31 July 2016. (a) Modeled integrated cloud water (g m^{-2}, 0-1000 m AGL) and (b) observed 30-min cloud-base fraction (%, 0-1000 m AGL) (Kalthoff et al., 2017, adapted) derived from ceilometer data with a temporal resolution of one minute. An observed cloud-base fraction of 100 % is related to 30 consecutive detections of low clouds in the 30 minute interval. The solid black lines separate the meteorological phases defined in Knippertz et al. (2017), including the Onset phase (Phase 1b) as hatched area. The focus time period for the case study simulation on D3 (3 July 2016) is framed by dashed lines in both panels.

Whereas in the *Onset phase* the stratus conditions are rather infrequent, visible in the observed and modeled results, the NLLS conditions in the *Post-onset phase* are more established. Especially the time period from 27 June to 12 July 2016 is characterized by extensive NLLS. 3 July shows especially intense NLLS (see dashed lines in Fig. 5.4) and was therefore identified as a golden-day by the DACCIWA community. Based on this decision, the present study focuses on 2-3 July 2016 with respect to domain D3.

The subsequent Figures 5.5, 5.6 and 5.7 present the COSMO-ART forecasts, averaged over the 57 h lead time from 13 June to 31 July 2016 (DACCIWA ground-based measurement campaign period), of the 2 m temperature, 2 m RH and the surface net downward shortwave radiation, respectively. For the 2 m temperature (Fig. 5.5) a significant cold bias of COSMO-ART on an order of magnitude of 4 K is visible, whereas during night the temperature is reasonably

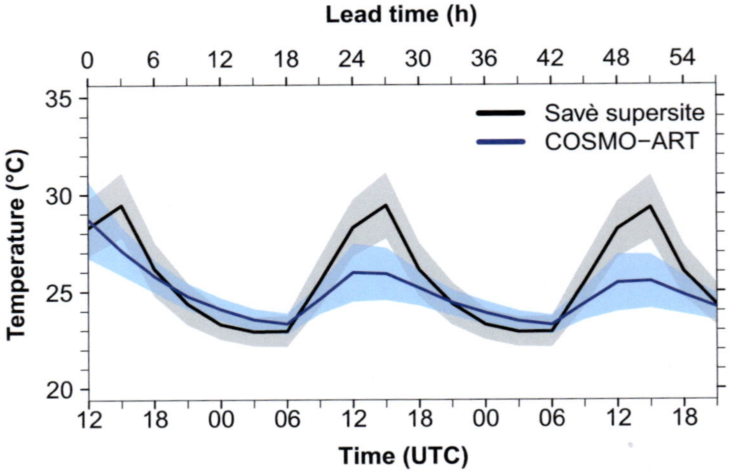

Figure 5.5: Temporal evolution of the 2 m temperature (°C) for the mean 57 h forecast lead time (13 June - 30 July 2016) at Savè as observed (black, Energy Balance Station) and modeled with COSMO-ART (blue). The shaded areas denote the standard deviation.

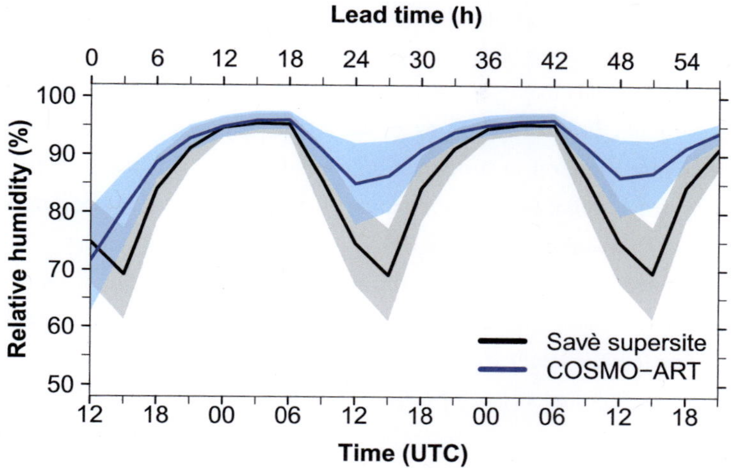

Figure 5.6: Same as Figure 5.5 but for 2 m RH (%).

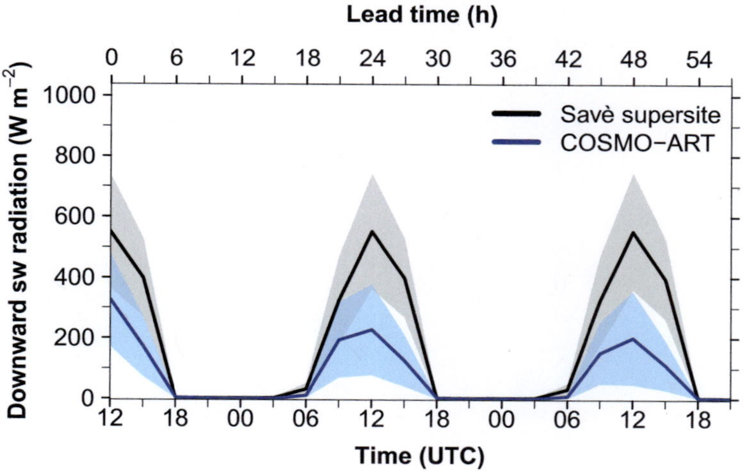

Figure 5.7: Same as Figure 5.5 but for surface net downward shortwave radiation (W m^{-2}).

reproduced. Linked with the cold bias in Figure 5.5, a wet bias in Figure 5.6 regarding RH is visible. Around 15 UTC, differences up to 15 % occur. The data analysis reveals that by using the convection parameterization of COSMO-ART, too much clouds in addition to the NLLS are produced in combination with an overestimation in precipitation (compare Fig. 5.12). As a result, the incoming solar radiation is significantly decreased (Fig. 5.7), leading to the deviations in the temperature and RH.

The identified shortcomings of the D1 setup in reproducing the SWA meteorological conditions lead to the decision not to use the D1 results for further case study analyses. For that purpose the setups D2 and D3 were developed and applied as described in the model setup description (Sect. 3.1.5).

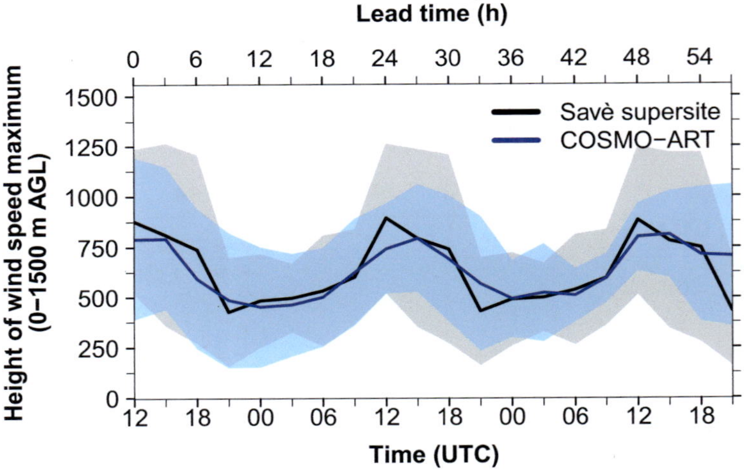

Figure 5.8: Temporal evolution of the height (m AGL) of the wind speed maximum between 0 and 1500 m AGL for the mean 57 h forecast lead time. (13 June - 30 July 2016) at Savè as observed (black, Doppler Lidar) and modeled with COSMO-ART (blue). The shaded areas denote the standard deviation.

Despite the deviations shown above, D1 reproduced the low-level wind and cloud conditions reasonably. Figure 5.8 indicates a good agreement between the observed and modeled diurnal cycle of the height of the low-level wind speed maximum at Savè. The NLLJ, that develops on average after 18 UTC and dissolves after 9 UTC and that is assumed to be of great importance for the formation of the NLLS (Adler et al., 2017), has its maximum at about 500 m AGL, increasing in height from 21 UTC to 9 UTC, in agreement with observations due to the development of the convective PBL. Further insight in the NLLJ characteristics is achieved by the mean diurnal cycle of the NLLJ over Savè in Figure 5.9, using the first 24 hours of the COSMO-ART forecasts, again averaged over the DACCIWA ground-based measurement campaign period. Displayed is the vertical profile of the wind speed for the lowest 2000 m ASL for COSMO-ART (Fig. 5.9a) and the observations (Fig. 5.9b). The lifting of the NLLJ is linked to the increase of turbulence in the PBL. Amplitude and height, the highest intensity occurs between 0 and 6 UTC, are well represented by the model. For the interpretation it has to be considered that the model output is 3 hourly and the observational data is averaged to hourly data. Figure 5.10 shows the mean cloud water concentration (g kg^{-1}, shaded colors) for the lowest 4 km ASL together with the mean ceilometer low cloud base detection (black bars).

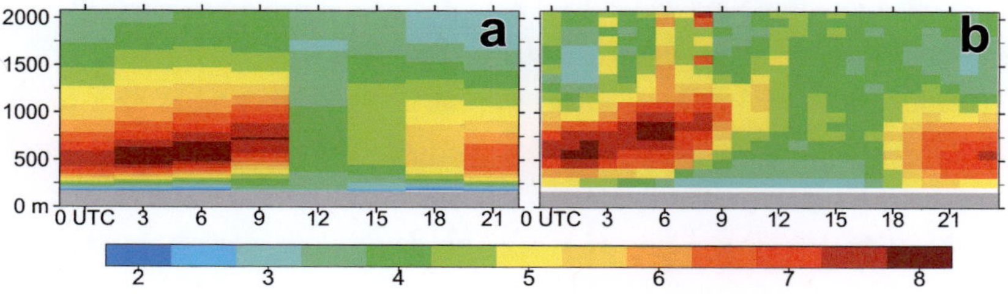

Figure 5.9: Wind speed profile (m s^{-1}) between 0 and 2000 m ASL as mean diurnal cycle (13 June - 30 July 2016) at Savè for (a) COSMO-ART and (b) Doppler Lidar observation.

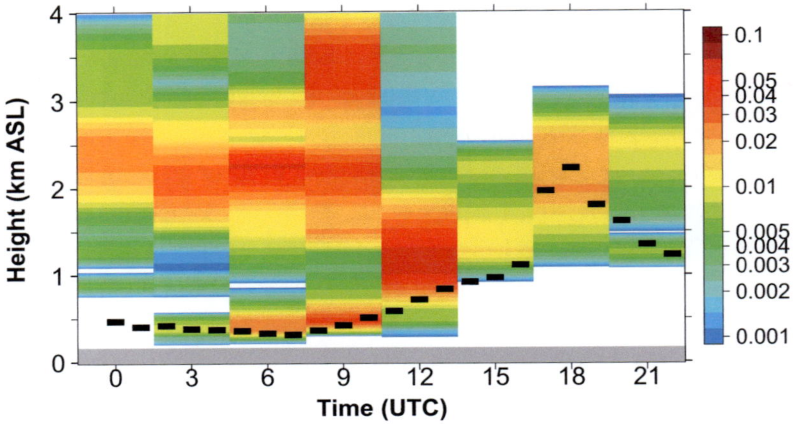

Figure 5.10: Modeled cloud water profile (g kg^{-1}) between 0 and 4 km ASL as mean diurnal cycle (13 June - 30 July 2016) at Savè. The black bars denote the temporal mean over the lowest ceilometer cloud base detection in the corresponding hour at Savè supersite.

The observations on average indicate a stratus development in the early morning with a cloud base below 500 m AGL and a lifting after 7 UTC. Significant cloud water in the model results is visible from 6 UTC on in a height that is comparable with the observations. Also the modeled clouds show a lifting after 7 UTC. Figure 5.10 therefore indicates a good model representation of the low cloud base height temporal evolution, even after the stratus transforms to cumuliform clouds in the afternoon.

The thickness of the low-level cloud layer can be assessed by the vertical profile of RH. During the intensive operation periods (IOP) Savè launched hourly soundings (frequent radiosoundings). Exemplarily this is shown for IOP 6 from 2 July 2016 21 UTC to 3 July 11 UTC. This day will also later be used for the case study analysis (Sect. 6). The observations in Figure 5.11a indicate the onset of a saturation at 22 UTC in a shallow layer around 250 m AGL. Based on the results of Figure 5.4b, the observed stratus layer is fully established at around 2:30 UTC.

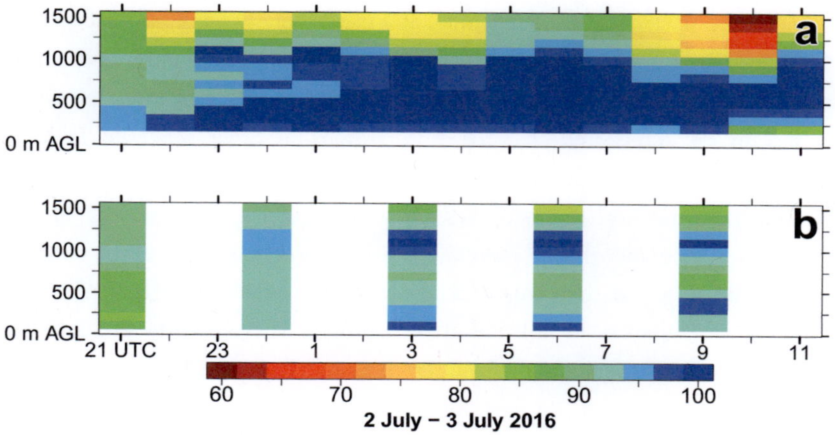

Figure 5.11: RH profile (%) between 2 July 2016 21 UTC and 3 July 2016 11 UTC (IOP 6) at Savè, (a) observed from frequent radiosounding and (b) COSMO-ART for the lowest 1500 m AGL.

The thickness of the saturated layer increases with time and between 2 UTC and 7 UTC the vertical extension covers the lowest 1 km AGL relatively constant. After 7 UTC the stratus layer thickness decreases. The comparison with the simulated stratus in Figure 5.11b is hindered by the 3 hourly ouput but it can be seen that saturation onset occurs 6 hours later at 3 UTC, which is on the other hand in good agreement with the observed time of full stratus establishment. The modeled cloud layer is restricted to the layer below 250-500 m AGL. Above this layer significant drier conditions (85-95% RH) occur. It can be concluded that for 3 July 2016 the time of the onset and dissolution of the stratus can be reasonably reproduced by the model whereas the vertical extension is significantly underestimated. Instead of forming a closed cloud layer in the lowest 1 km AGL, a double layer structure occurs, with one layer near the surface and one between 1000 and 1250 m AGL.
The daily precipitation amounts at Savè are displayed in Figure 5.12 for June and 5.13 for July 2016. This includes the modeled precipitation (red), the

Savè ombrometer measurement (blue) and the remote sensing product TRMM (TMPA-RT) (Huffman, 2016) that provides 3 hourly precipitation with a grid mesh size of 0.25 degrees (same grid mesh size as D1). For the comparison it has to be considered that the observations at Savè supersite start on 13 June 2016 and that the comparison between TRMM and COSMO-ART on the one hand and Savè supersite on the other hand is pixel to point comparison. The Savè supersite shows individual strong precipitation events between 20 and 60 mm and frequent weak precipitation events. COSMO-ART and TRMM precipitation events are more uniform and virtually at all times below 10 mm. The individual strong events observed from Savè supersite are not reproduced. While the two outstanding precipitation events in June on 16 and 19 June are located within the *Onset phase* (16-26 June), the *Post-onset phase* (22 June - 20 July) is characterized by a reduction in observed precipitation (Savè supersite and TRMM), whereas COSMO-ART still shows daily precipitation. Remarkable is the transition from the *Post-onset phase* to the *Wet westerly regime* (21-26 July) that is indicated by a shifting of wet conditions from tropical Atlantic into the Sahel (Knippertz et al., 2017). Significant higher precipitation amounts for all three datasets are visible. However, the strong precipitation events of Savè supersite are not reproduced by COSMO-ART and TRMM. One reason for this misrepresentation is located in the grid mesh size of 0.25° that is too coarse to reproduce very local precipitation peaks.

A statistical view on the precipitation time series is given in Table 5.1 as a general overview and Table 5.2, as a contingency table, also includes four skill scores. Both tables use the 48 day joint temporal coverage (12 June to 27 July and 29 to 30 July). Table 5.1 confirms the findings of the Figures 5.12 and 5.13. COSMO-ART shows the highest number of precipitation events (virtually every day). The average precipitation amount is on the same order of magnitude as the two observational products. Remarkable are the differences in the standard deviation.

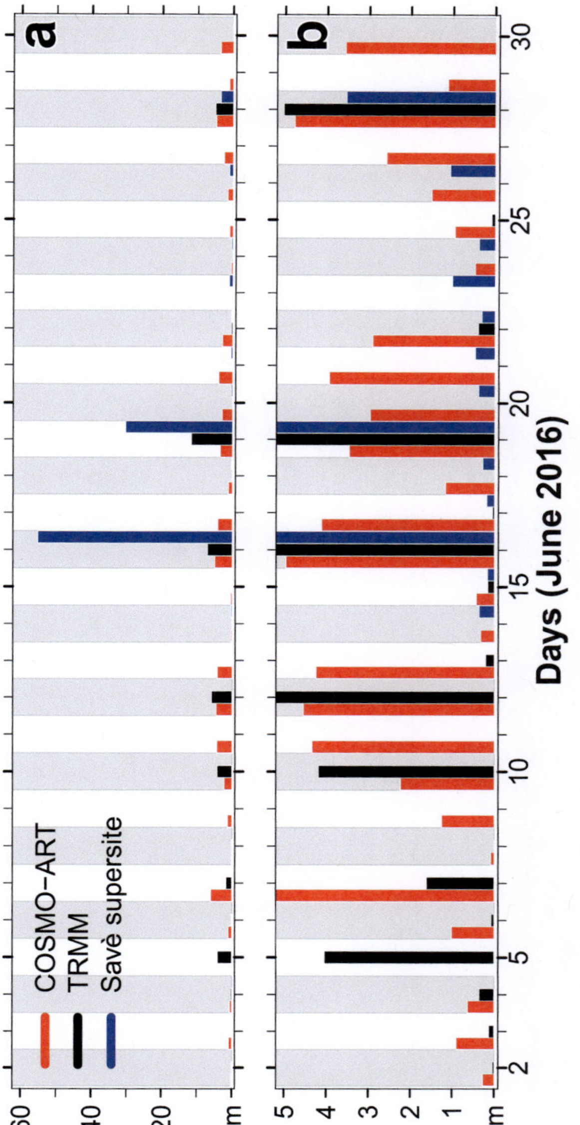

Figure 5.12: Daily precipitation (mm) for June 2016 of COSMO-ART (red), TRMM observation (black) and Savè supersite observation (blue). For better visibility, the figures (a) and (b) capture the precipitation range 0-60 mm and 0-5 mm, respectively. The gray areas separate single days. The observations at Savè supersite (blue) start on 13 June 2016.

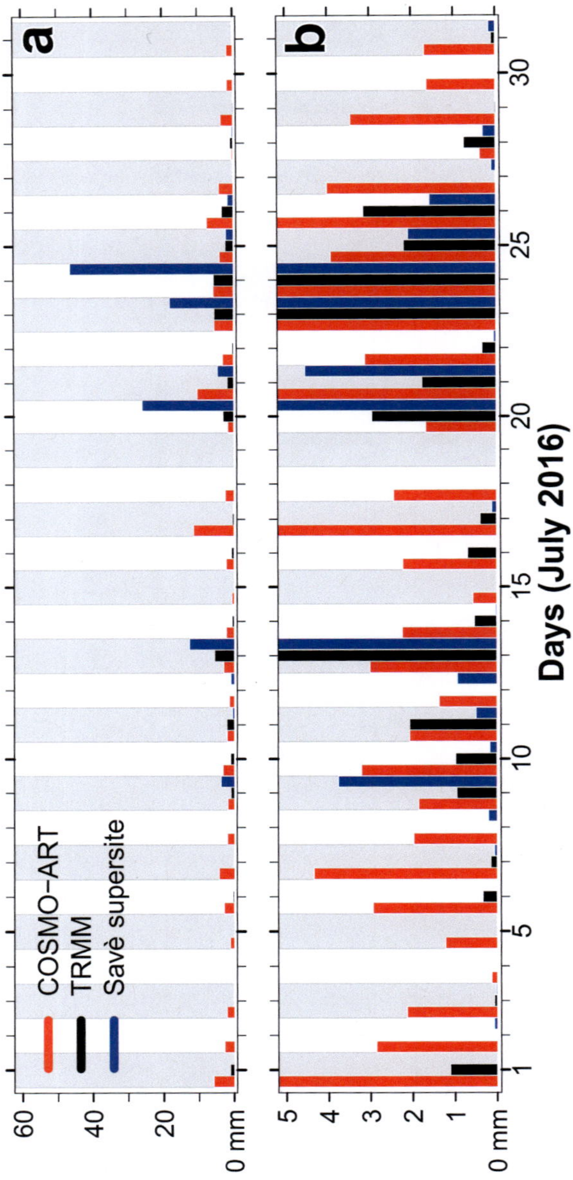

Figure 5.13: Same as Figure 5.12 but for July 2016.

Table 5.1: Savè precipitation quantified by COSMO-ART, TRMM and the Savè supersite for the 48 day joint temporal coverage within June-July 2016.

	COSMO-ART	TRMM	Savè supersite
Precipitation yes	47	28	38
Precipitation no	1	20	10
Daily average (mm)	2.9	1.2	4.4
Standard deviation	2.4	2.4	11.7

Table 5.2: Contingency table of the Savè precipitation quantified by COSMO-ART, TRMM and the Savè supersite for the 48 day joint temporal coverage within June-July 2016. *COSMO-ART/TRMM* evaluates COSMO-ART against TRMM, *COSMO-ART/Savè supersite* evaluates COSMO-ART against Savè supersite and *TRMM/Savè supersite* evaluates TRMM against Savè supersite. The skill scores *Thread Score* (TS, 1 means perfect), *Probability of False Detection* (POFD, 0 means perfect), *Peirce Skill Score* (PSS, 1 means perfect) and *Heidke Skill Score* (HSS, 1 means perfect) are added.

	COSMO-ART/ TRMM	COSMO-ART/ Savè supersite	TRMM/ Savè supersite
Hit	28	38	25
Missed	0	0	13
False alarm	19	9	3
Correct rejected	1	1	7
TS	0.60	0.81	0.61
POFD	0.95	0.90	0.30
PSS	0.05	0.10	0.36
HSS	0.06	0.15	0.26

COSMO-ART and TRMM show equal standard deviations but compared to Savè supersite the values are nearly 5 times lower. The high variations in precipitation are therefore neither captured by COSMO-ART nor by TRMM. The contingency table (Tab. 5.2) reveals the shortcomings in the simulated precipitation. In the first column, COSMO-ART is evaluated against TRMM and in the second column against Savè supersite. Additionally, TRMM is evaluated against Savè supersite in the third column. COSMO-ART hits all precipitation events of TRMM and Savè supersite but the number of false alarms is very high. TRMM, as expected, has a better skill compared to Savè supersite with less false alarms and more correct rejections but misses more than one third of the Savè supersite precipitation events. These results are reflected in the skill scores *Thread Score* (TS), *Probability of False Detection* (POFD), Peirce Skill Score (PSS) and Heidke Skill Score (HSS). The definitions are given in Appendix C. The TS focuses on the question how well the forecasted precipitation events correspond to the observed ones. In this case COSMO-ART in relation to Savè supersite shows highest skill. Regarding the POFD, that relates to the fraction of the observed non-precipitation events which were incorrectly forecasted as precipitation events, TRMM shows significant higher skill. The PSS is added to focus more on the question how well the forecast separates between "yes" and "no" events. The higher the PSS the higher the discrimination and a zero value means no discrimination. COSMO-ART shows nearly no discrimination, whereas the TRMM observations are partially able to discriminate the observed events at Savè supersite. In addition, the HSS provides information whether a forcast is better than a random forecast (greater than zero) or has the same skill as the random forecast (zero). The COSMO-ART forecasts show a low HSS of 0.06-0.15 and are therefore not significant better than a random forecast.
In summary, COSMO-ART suffer from false alarms and TRMM from missed events and their precipitation standard deviation is significantly smaller than the local Savè observations.

5.3.2 Meteorological Quantities at SWA

This section extends the forecast validation in terms of meteorology from the point observations at Savè supersite to the space of the entire SWA research domain, by using remote sensing products. Figure 5.14 opposes the simulated vertically integrated cloud water and the observed total cloud cover from

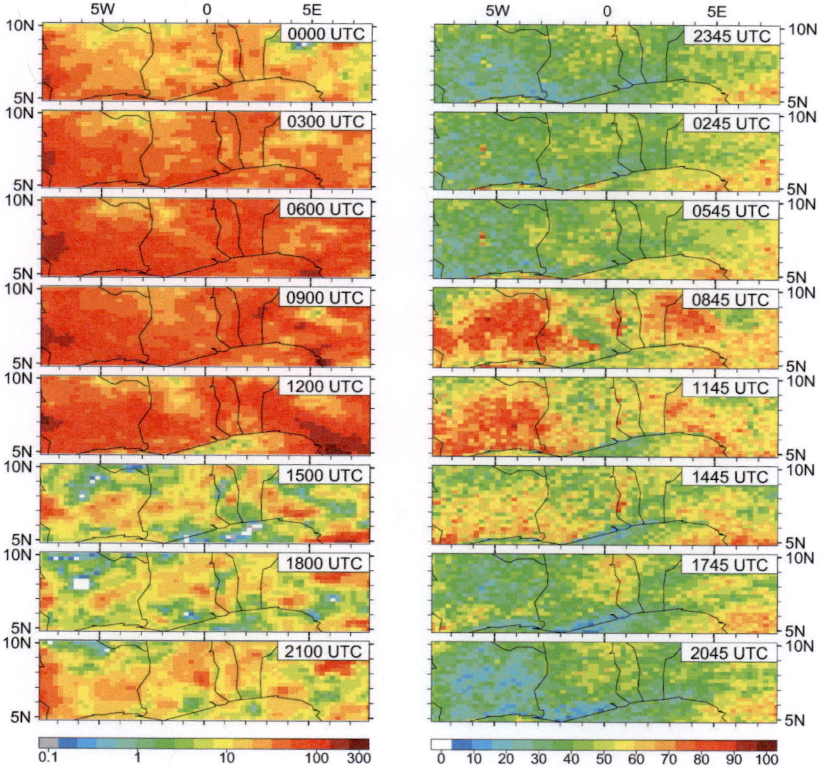

Figure 5.14: Mean diurnal cycle of modeled vertically integrated cloud water (g m^{-2}, left) and observed cloud cover from MSG/SEVIRI (%, right) averaged over the *Post-onset phase* (22 June - 20 July 2016).

MSG/SEVIRI (CMSAF, 2017), to qualitatively evaluate the mean location of clouds, especially the NLLS, during the *Post-onset phase*. As indicated by Figure 5.4, the *Post-onset phase* is characterized by a frequent occurrence of stratus. However, it is a well known problem that the MSG/SEVIRI product, and in general the remote sensing cloud products, have problems in capturing the low-level clouds during nighttime. This has to be considered in

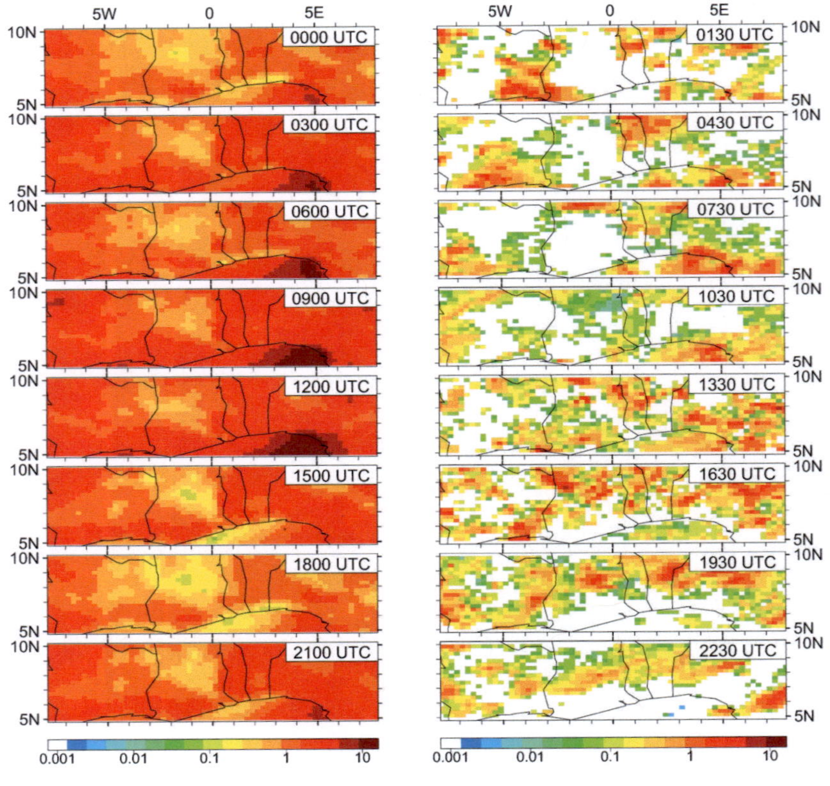

Figure 5.15: Mean diurnal cycle of precipitation (mm) in logarithmic scale, averaged from 2 June to 31 July 2016 for SWA, modeled by COSMO-ART (left) and observed by TRMM (right).

the interpretation of the data. Around 9 UTC the maximum of the NLLS is expected (Schuster et al., 2013) and observed at Savè during the DACCIWA campaign. In agreement with the observations, clouds are located over wide areas of Ivory Coast, the southweststern part of Ghana, over Benin and southern Nigeria. From 12 UTC to 15 UTC a clear reduction is visible in cloud water and cloud fraction. After 12 UTC the clouds are more in a zonal band around 7°N along the entire domain from Ivory Coast to Nigeria.
In Figure 5.15, the mean diurnal cycle of precipitation for June-July 2016 is shown as simulated by COSMO-ART (left) and observed by TRMM (right) in logarithmic scale. The TRMM product is the same as used for the comparison at Savè (Fig. 5.12 and 5.13). The differences between COSMO-ART and TRMM are remarkable. COSMO-ART shows on average more precipitation in terms of amount and area. However, the COSMO-ART results show similar precipitation patterns in the diurnal cycle, including less precipitation over Ghana and more precipitation east of Ghana. Highest precipitation amounts occur over the Gulf of Guinea southeast of Lagos, with a distinct diurnal cycle including the minimum at 18 UTC and the maximum at 12 UTC. In general, SWA shows lowest precipitation at 18 UTC. The observations also show enhanced precipitation in the eastern part of the Gulf of Guinea and comparably low precipitation over Ghana. In agreement between model and observation, precipitation occurs in the southern part of Ivory Coast and Nigeria, as well as in the more northerly parts of Togo and Benin.

5.3.3 Atmospheric Composition Quantities Locally at Savè Supersite

Aside from the meteorological conditions also the atmospheric composition can be validated against observations at Savè. Figure 5.16 shows the COSMO-ART near-surface PM10 concentration between 13 and 18 June 2016 together with the observations of the GRIMM device (black).

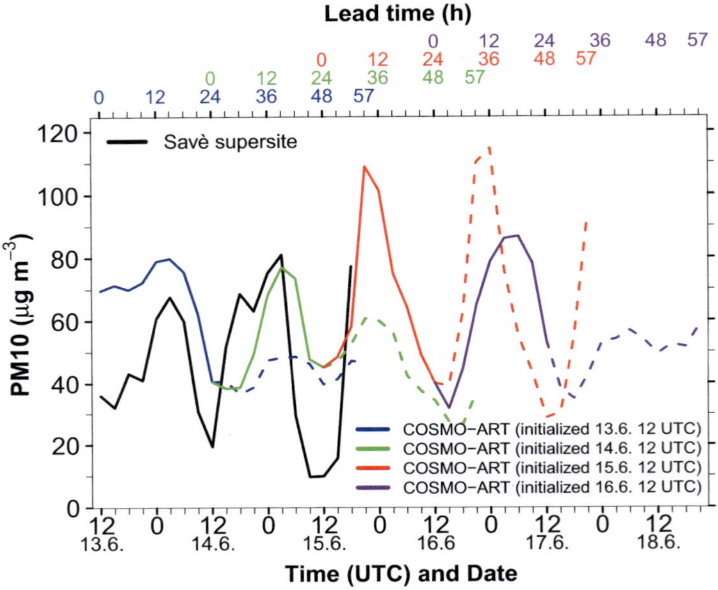

Figure 5.16: PM10 concentration (μg m^{-3}) between 13 June and 18 June, simulated with COSMO-ART for the four forecasts initialized on 13 (blue), 14 (green), 15 (red) and 16 June (purple) and observed at Savè supersite (black, GRIMM device). After 24 forecast hours (and with initialization of the next forecast) the model results are shown as dashed lines. The different lead times are denoted in the corresponding colors at the top abscissa. GRIMM fails after a few days of operation, therefore only the data of the first 2.5 days are shown.

As denoted in Section 5.1, the COSMO-ART forecasts were initialized every day at 12 UTC. This is indicated by the different colors in Figure 5.16. After 24 hours of simulation the lines are dashed to emphasize that a new simulation has been started at this time. The observations and modeled results show a distinct diurnal cycle of the PM10 concentration with highest values during night and lowest during daytime, likely related to the diurnal variation of the PBL depth. The nighttime maximum is well captured by COSMO-ART whereas the values at the daytime minimum are overestimated. Unfortunately, the data analysis is limited to this short time period of available observations.

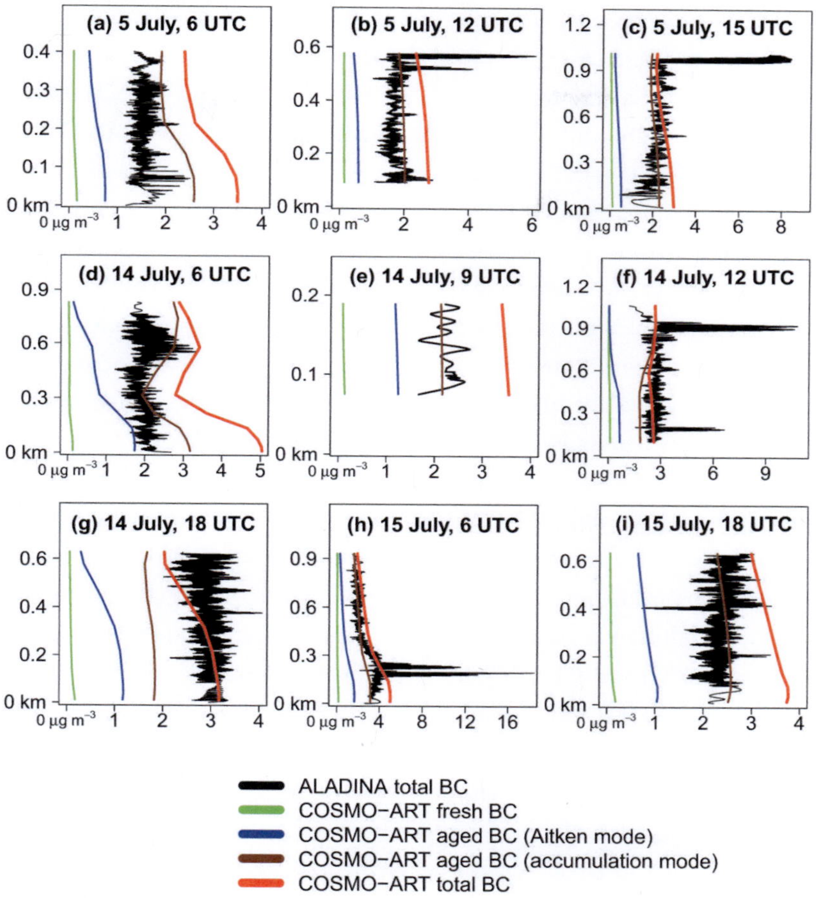

Figure 5.17: Vertical profiles (km AGL) of BC (μg m^{-3}) at Savè for (a-c) 5 July 2016, (d-g) 14 July 2016 and (h-i) 15 July 2016. The ALADINA observations of total BC are denoted in black, the COSMO-ART results for fresh BC, aged BC (Aitken mode), aged BC (accumulation mode) and total BC are shown in green, blue, brown and red, respectively. The observations were temporally assigned to the 3 hourly model output with a deviation not greater than 1 hour and by subsequently interpolating the model data to the ALADINA altitudes. Within these time steps, ALADINA conduct several ascends and descends. It is assumed that the observations within the time steps are measured instantaneously and the data is sorted according to their altitude, to allow for clearness of the visualization.

Further aerosol observations are provided by the unmanned aerial system ALADINA (Altstädter et al., 2015) that measured profiles of the BC concentration at Savè during the DACCIWA campaign. Figure 5.17 shows the spatiotemporally collocated total BC profiles (height AGL) of ALADINA (black) and COSMO-ART (red) for the three selected days 5, 14 and 15 July at 9 points in time. For COSMO-ART also the BC contributions from fresh BC (green), aged BC (Aitken mode, blue) and aged BC (accumulation mode, brown) are shown. This allows to estimate whether the BC comes from local sources or is advected and thereby aged. Riemer et al. (2004) indicate BC aging time scales of 8 hours close to the source regions and 2 hours above the source regions for daytime conditions during summer, predominantly related to the aging via the condensation of sulfuric acid on the BC particles.
On 5 July (Fig. 5.17a-c), the observations show a concentration maximum below 100 m AGL and lower, vertically constant BC concentrations above 100 m AGL in the morning. The COSMO-ART results also show higher concentrations in the near-surface layer up to 200 m AGL but overestimate the concentration by a factor of 2 to 2.5. At noon and in the afternoon the BC layers are vertically homogeneous with values around 2 μg m^{-3}. In the afternoon the COSMO-ART results match well with observations, especially in altitudes above 500 m AGL. Remarkable are the significantly increased BC concentrations in the observations at the top layers of 12 UTC and 15 UTC. There is ongoing work of the data producers to determine whether these values are artefacts due to clouds or can be verified. These strong signals are not reproduced by the model. For the 14 July (Fig. 5.17d-g) a full diurnal cycle of BC profiles from 6 UTC to 18 UTC is presented. In the morning hours the observations show increased concentrations above 500 m AGL with the peak at 600 m AGL. This peak is also well captured by the model. However, COSMO-ART shows near-surface concentrations of up to 5 μg m^{-3} that are 2.5 times higher than the observations. During noon the vertical profile is similar to this of the 5 July (Fig. 5.17b) with a vertically constant BC concentration.

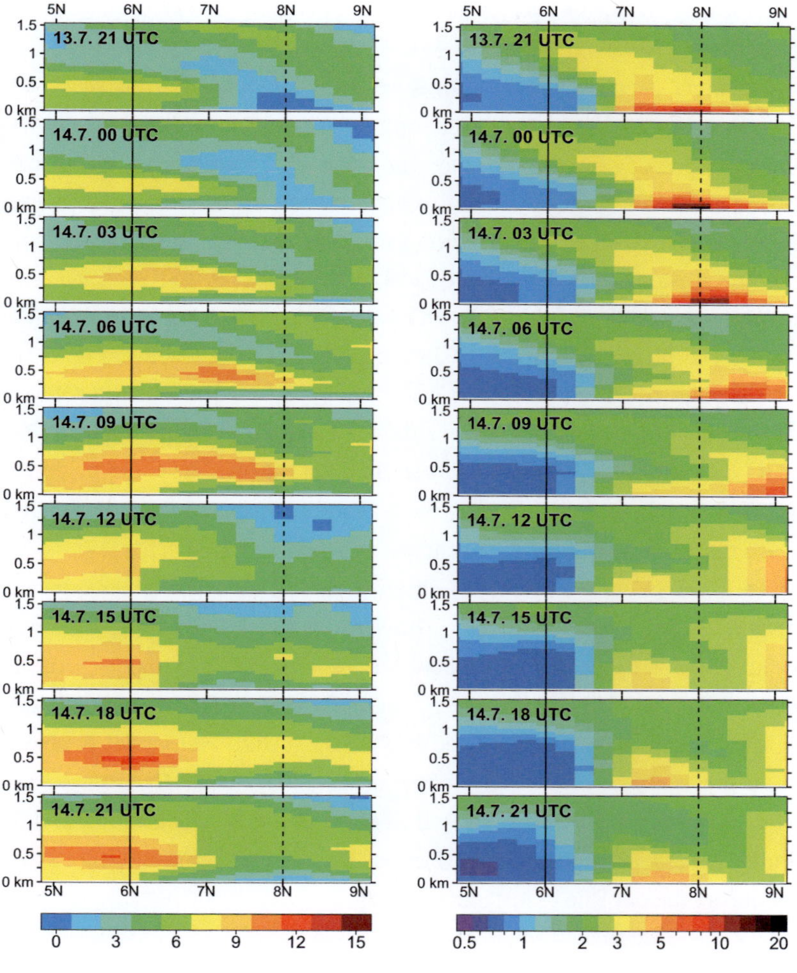

Figure 5.18: Meridional vertical transect (km AGL) of modeled (left) wind speed (m s^{-1}) and (right) total BC (μg m^{-3}) along the longitude of Savè from 13 July 21 UTC to 14 July 21 UTC. The dashed (solid) line indicates the location of Savè (the coast).

Furthermore, an observed BC concentration peak occurs in the top layer that is not seen by COSMO-ART. Between 12 UTC and 18 UTC no significant change in the vertical profile is visible.
The additional analysis of meridional vertical transects of wind speed and BC at Savè (Fig. 5.18) indicates a strong atmospheric flow from the Gulf of Guinea in northerly direction during night and in the morning. This airmass has comparable low BC concentrations but it pushes air with higher BC burden (affected from inland emissions during the day) northward. The wind speed maximum is around 500 m AGL, typical of the NLLJ. With the higher wind speed at 500 m AGL, a split of the BC airmass to a near-surface layer and an elevated layer occurs (compare 14 July 6 UTC in Fig. 5.17d and Fig. 5.18). Also the high observed peak at 12 UTC (Fig. 5.17f) fits to the COSMO-ART location of the elevated plume, although COSMO-ART cannot reproduce the BC concentration amplitude. These findings together with the negligible fraction of freshly emitted BC in Figure 5.17 indicate remote BC sources south of Savè rather than local Savè emissions.
At 6 UTC on 15 July (Fig. 5.17h) observation and model indicate enhanced concentrations below 400 m AGL. The vertical profile, capturing the lowest 900 m AGL, agrees very well between observation and model. Around 200 m AGL, ALADINA shows very high BC concentrations of more than 16 μg m^{-3} that are not represented in COSMO-ART. The model is able to capture the structure of the BC profiles observed by ALADINA. In general, model and observation show a vertical gradient in the BC concentration in the morning hours with higher near-surface values and a rather homogeneous BC layer during the day due to turbulent mixing.
The site at Savè was also equipped with ground-based devices to measure the volume mixing ratio (ppbv) of CO (Fig. 5.19) and O_3 (Fig. 5.20). As for the Figures 5.5-5.8, the mean 57 h forecast lead time is shown with the model realization in blue and the observations in black.

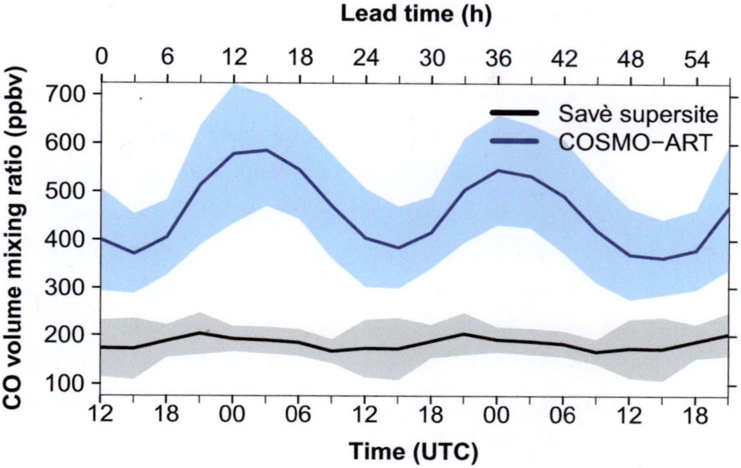

Figure 5.19: Temporal evolution of CO volume mixing ratio (ppbv) for the mean 57 h forecast lead time (13 June - 30 July 2016) at Savè as observed (black) and modeled with COSMO-ART (blue). The shaded areas denote the standard deviation.

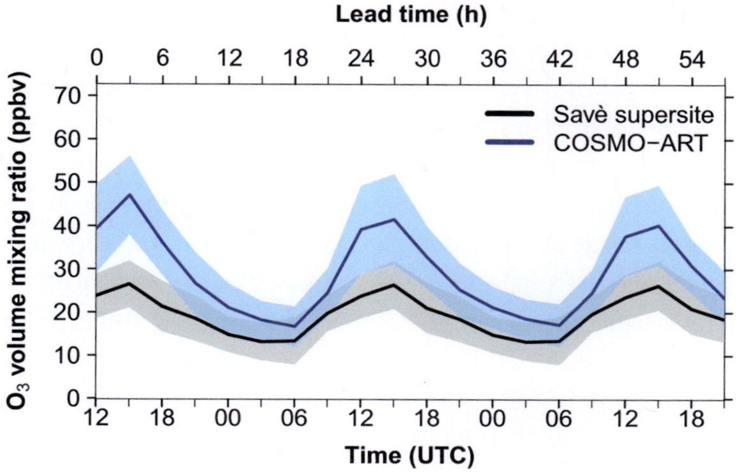

Figure 5.20: Same as Figure 5.19 but for O_3 volume mixing ratio (ppbv).

The CO concentration is seriously overestimated by COSMO-ART. Regarding O_3 a better representation can be achieved but also in this case COSMO-ART overestimates the concentrations. The diurnal cycle follows the findings of Figure 5.16. It is supposed that the anthropogenic emission of CO are overestimated. Within a sensitivity study (not shown) the CO emissions were reduced by two third in the entire domain. Since a CO reduction leads to a reduction in CO oxidation and therefore a reduced formation of hydro-peroxy radicals as precursors for the tropospheric O_3 formation, also the O_3 concentration was reduced in better agreement with observations.

The AOD time series for Savè is shown in blue in Figure 5.21, separated in the aerosol components sea salt (Fig. 5.21a), mineral dust (Fig. 5.21b), anthropogenic aerosol (Fig. 5.21c) and total aerosol (Fig. 5.21d). This is complemented by the AOD of the operational global chemistry model CAMS (green) available from CAMS (2017b) and the derived AOD from the sun photometer in Savè (red) available from AERONET (2017). Analogue to Figure 5.21, the sites Ilorin (Nigeria) (Fig. 5.22) and Koforidua (Ghana) (Fig. 5.23) are presented, for which also AERONET observations are available. The location of the three sites are given in Figure 2.1.

Although the mineral dust influence on SWA is reduced during the monsoon time, it significantly contributes to the total AOD. COSMO-ART shows very good agreement of the amplitude and temporal evolution of the mineral dust AOD compared to CAMS (Fig. 5.21a). This can be seen as a proof of the performance of the mineral dust emission parameterization that was newly implemented within this study, as described in Section 3.1.3. The AOD contribution of sea salt is one order of magnitude smaller compared to mineral dust (Fig. 5.21b). For sea salt, COSMO-ART shows a positive bias compared to CAMS. Highest differences and simultaneously highest uncertainty occurs for the anthropogenic aerosol (Fig. 5.21c). The COSMO-ART AOD for anthropogenic aerosol is three times higher than in CAMS. This overestimation is generally confirmed when considering the total AOD in Figure 5.21d in which also the AERONET AOD is added.

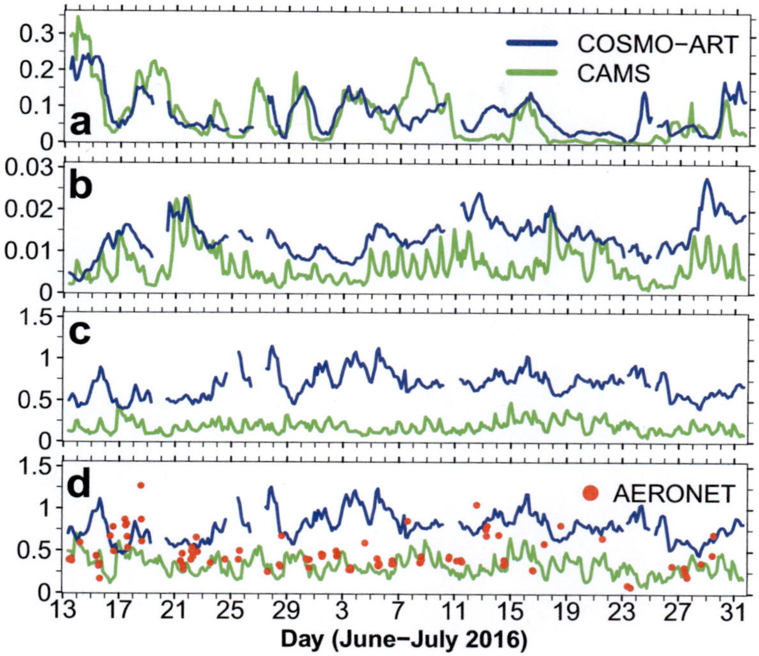

Figure 5.21: AOD (550 nm) at Savè from COSMO-ART (blue), CAMS (green) and AERONET (red) between 13 June - 31 July 2016 for (a) mineral dust, (b) sea salt, (c) anthropogenic aerosol and (d) total aerosol. Consider the different scaling of the ordinates. Data gaps are related to technical issues during the forecast.

The CAMS results in principle correspond to AERONET, also due to the fact that CAMS assimilates the AERONET AOD in contrast to COSMO-ART. The uncertainty in the anthropogenic emissions is assumed to be high and therefore also for the AOD comparison the anthropogenic aerosol AOD shows greater deviations than the natural aerosol.

For Ilorin in Nigeria the AOD overview (Fig. 5.22) looks similar to Savè with regard to the natural aerosol, with a very good representation of the mineral dust and small positive bias in the sea salt.

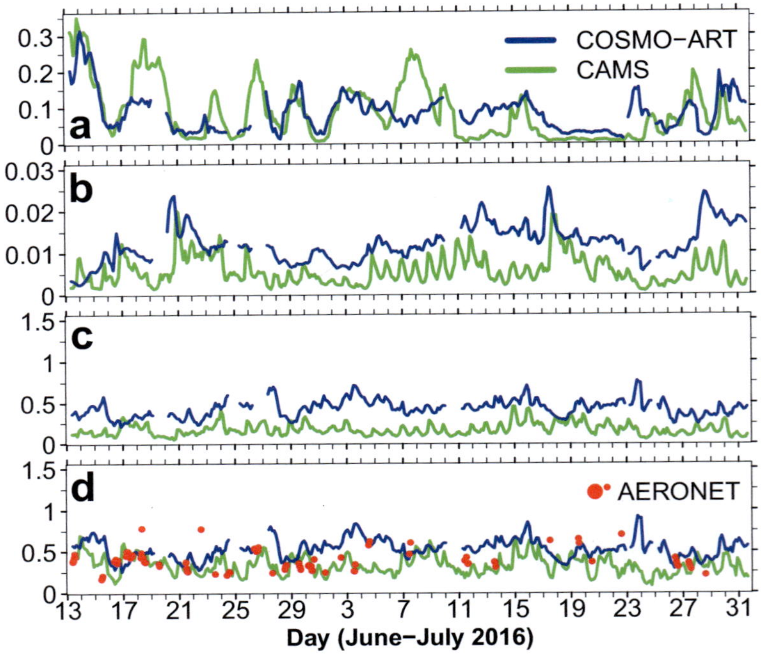

Figure 5.22: Same as Figure 5.21 but for Ilorin (Nigeria). See Figure 2.1 for location.

However, the COSMO-ART overestimation of anthropogenic aerosol is much smaller for Ilorin and the intercomparison with AERONET shows significantly better agreement. In contrast to Savè, that is affected by the Lomé and Accra city plumes in COSMO-ART, Ilorin seems not to be affected by the Lagos city plume.

Koforidua (Fig. 5.23) is the site nearest to the Gulf of Guinea. Therefore the mineral dust contribution is reduced compared to Savè and Ilorin (Fig. 5.23a) and the sea salt influence is increased (Fig. 5.23b). The anthropogenic aerosol again seems to be overestimated by COSMO-ART when comparing with CAMS (Fig. 5.23c).

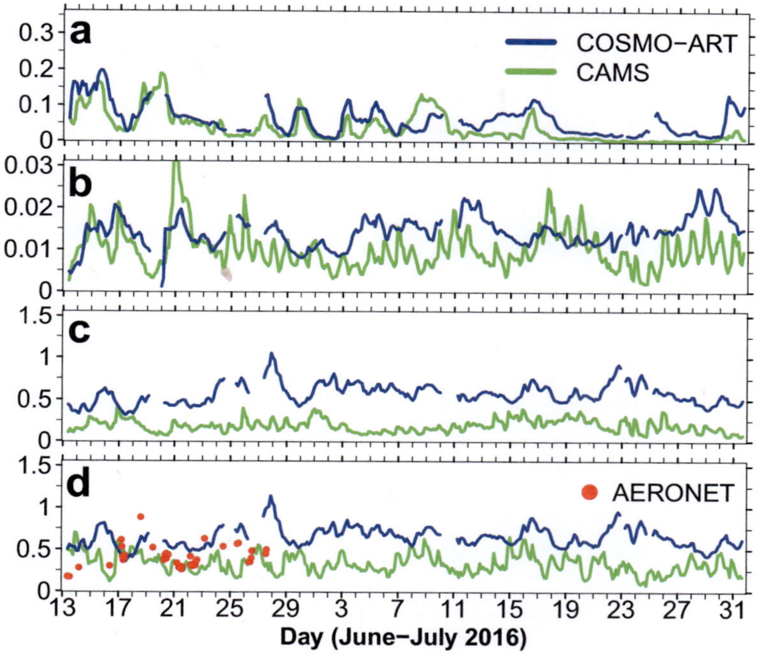

Figure 5.23: Same as Figure 5.21 but for Koforidua (Ghana). See Figure 2.1 for location.

Nevertheless, the available AERONET observations indicate reasonable results for the COSMO-ART total AOD (Fig. 5.23d). To allow for a comparison of the COSMO-ART AOD between the three sites, Figure 5.24 summarizes the results. The mineral dust concentration is zonally symmetrical with lower values to the South and higher values to the North in the vicinity of the Sahara. The dust is transported from northeast to southwest. These two aspects are reflected in Figure 5.24a. Savè and Ilorin are the northernmost sites with similar latitudes. Here we find the highest mineral dust AOD. Koforidua is the southernmost site and has therefore the lowest mineral dust AOD. The differences are on an order of magnitude of 0.1.

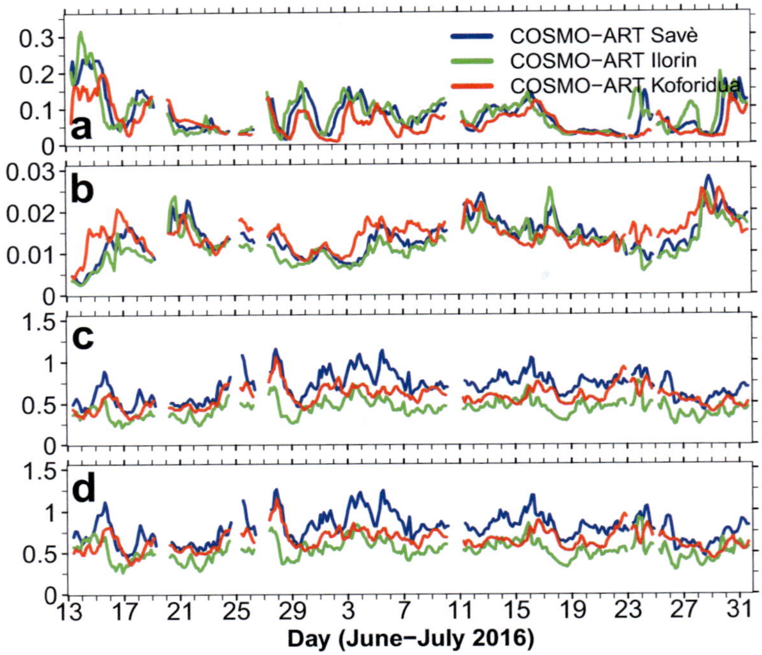

Figure 5.24: Simulated AOD (550 nm) at Savè (blue), Ilorin (green) and Koforidua (red) between 13 June - 31 July 2016 for (a) mineral dust, (b) sea salt, (c) anthropogenic aerosol and (d) total aerosol, summarizing the COSMO-ART AOD timeseries of Figure 5.21-5.23. Consider the different scaling of the ordinates. Data gaps are related to technical issues during the forecast.

Furthermore, a time lag between the three sites is obvious. The mineral dust reaches Ilorin (easternmost site) at first and Koforidua (westernmost site) at last. The time lag between Ilorin and Koforidua is approximately 1 day. This rather long traveling time is due to the fact that it is not only a transport from east to west but from northeast to southwest, so the dust plume also has to extend to the South to reach Koforidua. For sea salt (Fig. 5.24b), a contrary behaviour is visible with decreasing AOD with increasing distance from the coast. In this case no time lag is visible. Regarding the anthropogenic aerosol

(Fig. 5.24c), Savè shows a higher AOD than Ilorin, likely because Savè is located inside the Accra and Lomé plumes whereas Ilorin is not located in the Lagos plume. Since Lagos shows the highest city emissions of the entire research domain (not shown), a location within the plume of Lagos would certainly lead to highest AOD values. The comparison of the total AOD is shown in Figure 5.24d. The differences between Savè with the highest AOD and Ilorin with lowest AOD values is on the order of magnitude of 0.25.

5.3.4 Atmospheric Composition Quantities at SWA

Figure 5.25 shows the comparison of the total AOD at 550 nm simulated by COSMO-ART (Fig. 5.25a,c) and observed by MODIS Terra (Fig. 5.25b) and MODIS Aqua (Fig. 5.25d) regarding the ASOP. The COSMO-ART AOD is spatiotemporally collocated with the MODIS observations to enable a comparison. Additionally, in Fig. 5.26 the number of observations that are available for MODIS, and the differences between COSMO-ART and MODIS Terra and Aqua are shown in Figure 5.27.

The number of observations in Figure 5.26 already indicate serious limitations over land due to the frequent cloud cover that is even worse for Terra (Fig. 5.26a) than for Aqua (Fig. 5.26b). Over water, excluding the high precipitation area south of Nigeria, the data density is significantly improved. Therefore the results over water are supposed to be more robust.

In the MODIS observations the biomass burning emission plume, coming from the East and propagating as a pollution tongue in westerly direction over the Gulf of Guinea, is clearly visible. For COSMO-ART this pollution tongue is in good agreement when comparing with Aqua (Fig. 5.25c), but less pronounced in comparison to Terra (Fig. 5.25a). Over land the COSMO-ART AOD is significantly higher compared to MODIS. This is likely related to the anthropogenic aerosol and supports the findings regarding the comparison with AERONET and CAMS (Fig. 5.21d, 5.22d and 5.23d).

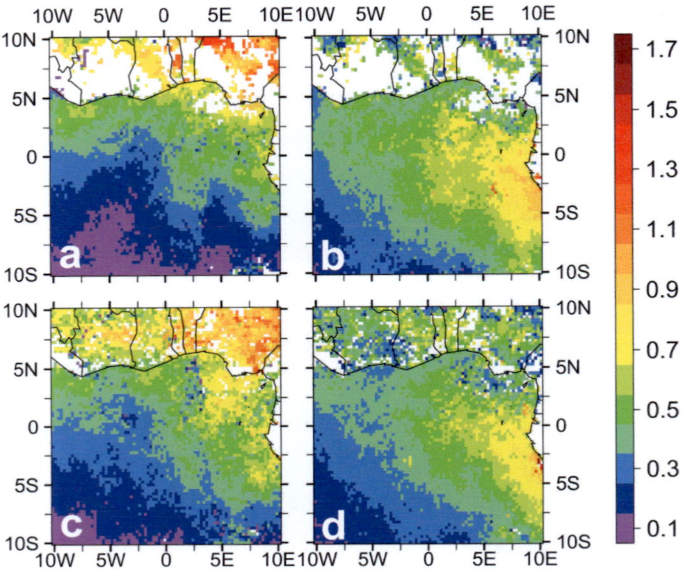

Figure 5.25: Mean total AOD averaged from 27 June to 17 July 2016 (ASOP) of (a) COSMO-ART, spatiotemporally collocated with MODIS Terra, (b) MODIS Terra, (c) COSMO-ART, collocated spatiotemporally with MODIS Aqua and (d) MODIS Aqua.

The AOD difference between COSMO-ART and MODIS, as visible in Figure 5.27, reveals an underestimation in the biomass burning plume over water between 0.2 and 0.5 and an overestimation in coastal areas of 0.1-0.3 and even higher overestimations over Nigeria.

The forecast validation section closes with the comparison of the COSMO-ART results with selected observations of the research aircraft Twin Otter, considering the atmospheric trace gas composition. Two 30 minute time slices around 9 UTC at 10 July (Fig. 5.28, TO-11) and 14 July (Fig. 5.29, TO-16) were selected. Both are located downstream of Lomé in a comparable height of 500-750 m AGL on average. TO-11 is directed along and TO-16 perpendicular to the atmospheric flow.

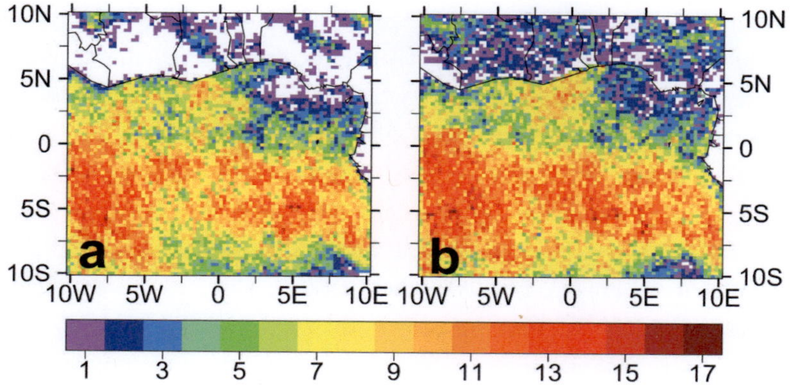

Figure 5.26: Number of observations within the time period 27 June - 17 July (ASOP) of (a) MODIS Terra and (b) MODIS Aqua

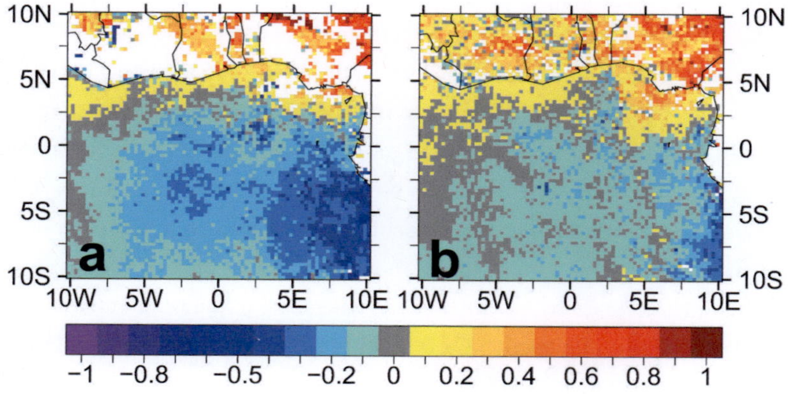

Figure 5.27: Mean AOD difference (27 June - 17 July, ASOP) between (a) COSMO-ART and MODIS Terra and (b) COSMO-ART and MODIS Aqua

Figure 5.28e and 5.28f show the temperature and specific humidity for TO-11, respectively. COSMO-ART (blue) reveals a cold and dry bias compared to the observation (black). The underestimation of the temperature is already indicated by the comparison with Savè (compare Fig. 5.5). The observed concentrations of the trace gases (Fig. 5.28e-l) are elevated in the vicinity of Lomé and stabilize at a lower level with increasing distance. COSMO-ART very well captures the higher values near Lomé when focusing on CO (Fig. 5.28g), NO_2 (Fig. 5.28i) and NO_x (Fig. 5.28j). O_3 seems not to be affected by the Lomé city emissions and is rather constant. The SO_2 concentration is very low in COSMO-ART and the observations.
COSMO-ART tends to slightly overestimate the trace gas concentrations with increasing distance from Lomé. Figure 5.28c and 5.28d indicate the approach of the aircraft to the pollution plume, likely related to Accra. This is visible in the modeled trace gas concentration increase. On the contrary, the observations do not confirm the encounter of a pollution plume, that is especially obvious in Figure 5.28d. Furthermore, Figure 5.28d shows an intrusion of comparable clean air with the monsoon flow that can penetrate approximately 20 km inland.
Figure 5.29 focuses on a flight, perpendicular to the downstream flow of Lomé. Also in this case COSMO-ART shows the Accra plume in the horizontal distribution (Fig. 5.29c) and the vertical transect (Fig. 5.29d). The observations generally show an increase in the trace gas concentrations when flying at lower heights, that is especially evident for NO, NO_2 and NO_x (Fig. 5.29h-j), but no distinct Accra plume when approaching to the northwesterly edge of the flight track. The second NO_x peak around 9 UTC is very well captured in location and amplitude by COSMO-ART, whereas the first peak at 08:50 UTC is not represented by the model. O_3 is slightly overestimated by COSMO-ART and CO is overestimated by a factor of 2. The vertical gradient in the NO_x concentration between 500 m AGL and 1250 m AGL between 08:55 UTC and 09:00 UTC is well reproduced by COSMO-ART but generally the observed pollution layer seems to be more shallow than the modeled one (Fig. 5.29d).

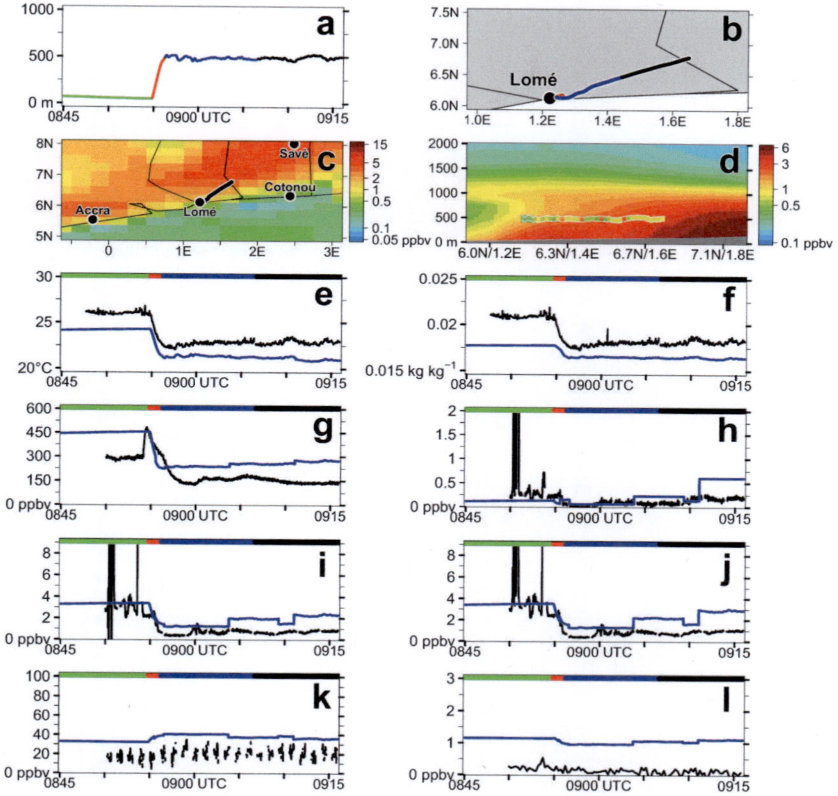

Figure 5.28: Comparison of Twin Otter measurement flight TO-11 (10 July 2016, 08:44 UTC to 11:37 UTC) results with COSMO-ART. For a comparison, the model output of 9 UTC and the measurements 15 minutes around this time step (08:45-09:15 UTC) were selected. (a) Flight altitude (m AGL), (b) flight track, (c) NO_x concentration (ppbv) at 500 m AGL and flight track, (d) vertical transect of NO_x concentration (ppbv) along the flight track with aircraft observations included, (e) temperature (°C), (f) specific humidity (kg kg^{-1}), (g) CO concentration (ppbv), (h) NO concentration (ppbv), (i) NO_2 concentration (ppbv), (j) NO_x concentration, (k) O_3 concentration (ppbv) and (l) SO_2 concentration (ppbv). The panels (e)-(l) present the COSMO-ART results in blue and the observations in black. The horizontal color lines on top of these panels refer to the colors in panel (a) and (b) to illustrate the aircraft location related to the observed trace gas concentrations. The vertical transect in panel (d) only captures the flight track at 500 m AGL and not the hook-shaped flight pattern near Lomé.

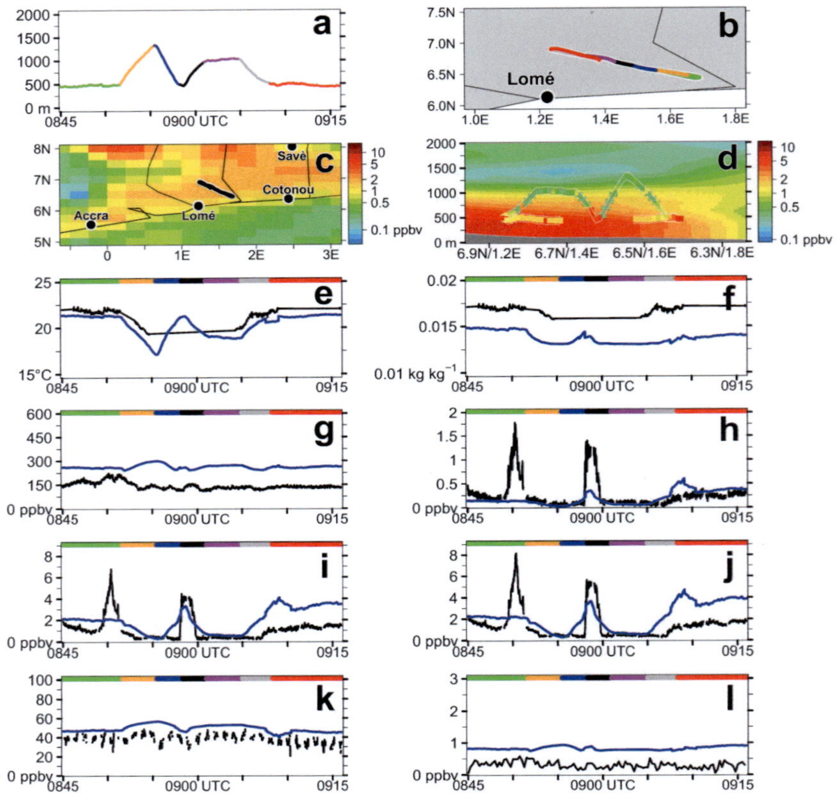

Figure 5.29: Comparison of Twin Otter measurement flight TO-16 (14 July 2016, 06:44 UTC to 09:50 UTC) results with COSMO-ART. For a comparison the model output of 9 UTC and the measurements 15 minutes around this time step (08:45-09:15 UTC) were selected. (a) Flight altitude (m AGL), (b) flight track, (c) NO_x concentration (ppbv) at 750m height and flight track, (d) vertical transect of NO_x concentration (ppbv) along the flight track with aircraft observations included, (e) temperature (°C), (f) specific humidity (kg kg^{-1}), (g) CO concentration (ppbv), (h) NO concentration (ppbv), (i) NO_2 concentration (ppbv), (j) NO_x concentration, (k) O_3 concentration (ppbv) and (l) SO_2 concentration (ppbv). The panels (e)-(l) present the COSMO-ART results in blue and the observations in black. The horizontal color lines on top of these panels are denoted to the colors in panel (a) and (b) to illustrate the aircraft location related to the observed trace gas concentrations. The vertical transect in panel (d) captures the full 30 minute flight track in contrast to Figure 5.28.

5.3.5 Forecast Validation Summary

This section has compiled a large number of comparisons between COSMO-ART and available ground-based, airborne and remote sensing observational datasets to assess the performance of the COSMO-ART forecasts conducted during the DACCIWA campaign in June-July 2016. The intercomparison includes meteorological parameters as well as the atmospheric composition including trace gases and aerosols. It has to be considered that, aside from the remote sensing data, the model data (with a rather coarse grid mesh size of 28 km) is compared to point observations.

The COSMO-ART forecasts are able to reasonably reproduce the NLLS characteristics including the stratus occurrence in the different meteorological phases during the DACCIWA campaign, the spatial distribution, diurnal cycle, cloud base and NLLJ. Regarding the basic meteorology, COSMO-ART reveals shortcomings which appear in an underestimation of the daytime temperature (about 2 K) and incoming shortwave radiation (200-300 W m^{-2}) linked with an overestimation of cloud cover. The precipitation forecasts show low sharpness, so cannot reproduce the strong events observed by the Savè supersite, and have a high probability of false detection.

Concerning the atmospheric composition, COSMO-ART captures the features of the research domain, including the pollution from the cities and the biomass burning plume. Even without assimilation of observations, COSMO-ART shows reasonable agreement regarding the vertical profiles of BC over Savè, the temporal evolution of the mineral dust AOD and in general for the natural aerosol with regard to CAMS. A general tendency of COSMO-ART to overestimate the air pollution from anthropogenic origin can be observed when comparing with AERONET, MODIS and CAMS. The anthropogenic aerosol AOD modeled by COSMO-ART at Savè is on average three times higher than in CAMS. On the other hand, the COSMO-ART forecasts show good agreement regarding aircraft observations of trace gases in which even narrow peaks can be captured.

Several sources of uncertainty might contribute to the disagreements between COSMO-ART and the observations. This includes the model with its meteorological and aerosol/chemistry part and the parameterizations, the initial and boundary data including meteorology and the atmospheric pollution emission inventories as well as the observations itself. In Section 6.1.1 it will be shown that a significant portion of the deviations in the meteorological results are related to the specification of the forecast model setup D1. Significant improvements are reached via D2. This includes a finer grid mesh size (5 km instead of 28 km) as well as the use of the TMMS and explicit convection. The parameterized convection is identified as the major issue leading to an overestimation in cloud cover and precipitation and an underestimation in incoming solar radiation related to a cold bias. The meteorological boundary data of ICON is very likely not responsible for the disagreement since it does not exhibit a cold bias. By considering the studies of Marsham et al. (2013) and Pearson et al. (2013), the use of explicit convection even in combination with a coarse grid mesh size would likely have improved the meteorological forecast performance.

Regarding aerosols and trace gases the major issue is addressed to the uncertainty in the emission inventories including anthropogenic and biomass burning pollution. This intercomparison study is performed during the early phase of the DACCIWA processing and supply of aerosol and trace gas observational data. Therefore further investigation, by using the full capacity of the DACCIWA observations and also intercomparison with further DACCIWA-related chemistry modelling, is necessary.

The detailed assessment of the uncertainty in the observations goes beyond the scope of this study. However, in terms of remote sensing products it is very likely that their reliability suffer from limited data coverage through the frequent cloud cover over SWA.

The forecast realization over a long time period together with the performance confirm the applicability of COSMO-ART as an operational aerosol/chemistry forecast model system.

6 Aerosol Feedback Case Study Simulations

The realization and analysis of aerosol feedback simulations for a case study focusing on 2-3 July 2016 is the focal point of this work, beside the COSMO-ART forecast realization as presented in Section 5. Scientific discussions in the DACCIWA working groups Cloud-Aerosol Interactions and Boundary Layer Dynamics identified 3 July as one of the golden days, suitable for further investigation. This day was characterized by especially intensive and persistent stratus, as already indicated by Figure 5.4. Furthermore, 3 July is located in the center of the *Post-onset phase* and it is expected that the undisturbed monsoon conditions in this phase favor and support the process studies. Since the meteorological conditions show less variation from day to day, it is assumed that, even with a focus on a very short time period, insight can be achieved that can be generalized to virtually the length of the *Post-onset phase*.

6.1 Modeling Strategy

The following section describes the strategy for realizing the COSMO-ART aerosol feedback simulations. This includes the realization at domain D2, with the grid mesh size of 5 km, as intermediate step in the increase of the grid mesh size between ICON (approximately 13 km) and the target grid mesh size of 2.5 km (domain D3). Domain D3 is used for the model realizations that consider the aerosol feedbacks. The general specifications of D2 and D3 are given in Section 3.1.5.

6.1.1 Intermediate Modeling Domain (D2)

Due to the meteorological shortcomings of COSMO-ART with the D1 setup (Sect. 5), we refrain from realizing the case study by nesting into the forecast modeling domain D1. Instead, the simulation setup D2 (Tab. 3.7) was applied. D2 also uses ICON as meteorological boundary conditions but dynamically scales down to 5 km by using the TMMS without prognostic aerosol. For D2, the convection parameterization was switched off, which significantly improved the model performance e.g. in terms of radiation, temperature and RH (Fig. 6.1-6.3). By using D2, a nine-day period was simulated to allow for a spin-up of the aerosol and trace gas temporal evolution. To capture all relevant pollution sources (e.g. biomass burning from central Africa), the simulation domain size spans an area similar to that of D1 (Fig. 3.15) with 1000 times 1000 grid points.

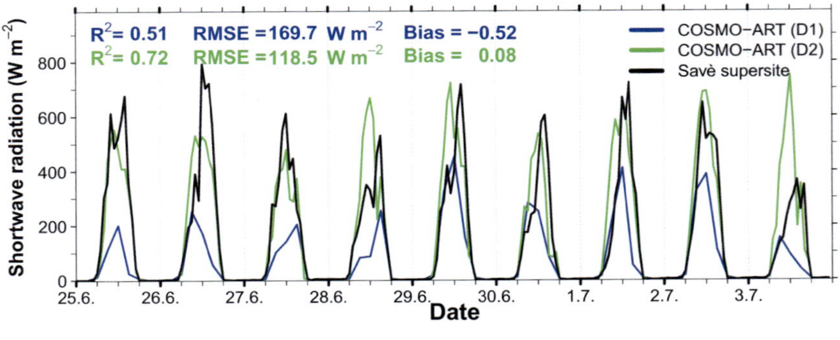

Figure 6.1: Temporal evolution of the surface net downward shortwave radiation (W m^{-2}) for the nine-day spin-up time (25 June - 3 July 2016) at Savè as observed (black, Energy Balance Station) and modeled with COSMO-ART (D1 in blue and D2 in green).

Figure 6.1 shows the comparison of the surface net downward shortwave radiation at Savè between 25 June and 3 July regarding D2 (green) and the observation (black). The COSMO-ART results of D1 (blue) as well as the coefficient of determination (R^2), Root Mean Square Error (RMSE) and the bias are added to emphasize the improvement in D2 with respect to D1. With D2 the diurnal cycle together with the daytime maximum is significantly better represented than in D1. Related to this improvement, better performances in the 2 m temperature (Fig. 6.2) and RH (Fig. 6.3) can also be observed.

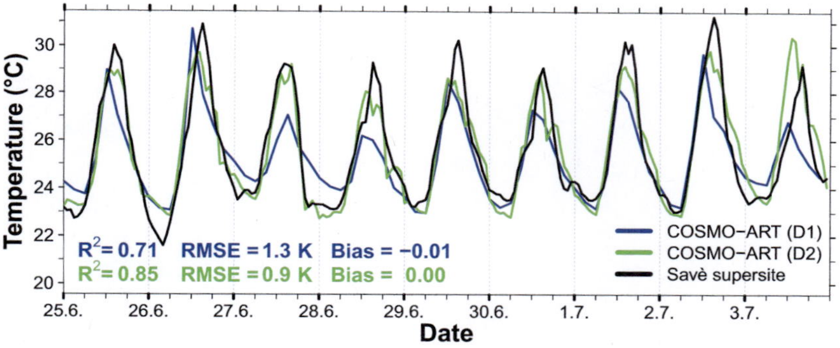

Figure 6.2: Same as Figure 6.1 but for 2 m temperature (°C).

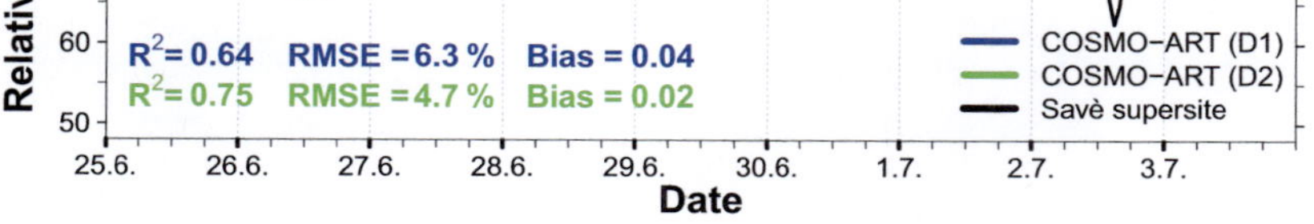

Figure 6.3: Same as Figure 6.1 but for 2 m RH (%).

Within D2 the RMSE of the temperature can be reduced below 1 K, with a better representation of the daytime maximum and the nighttime minimum. Nevertheless, at certain days a cold bias of the daytime maximum of 1-2 K remains. This leads to a wet bias in RH on these days (see Fig. 6.3). Due to the better nighttime temperature representation in D2 the nighttime wet bias vanishes.

As done for D1 in Figure 5.10, Figure 6.4 shows the diurnal cycle of the modeled cloud water profile for D2 at Savè together with the ceilometer cloud base observations.

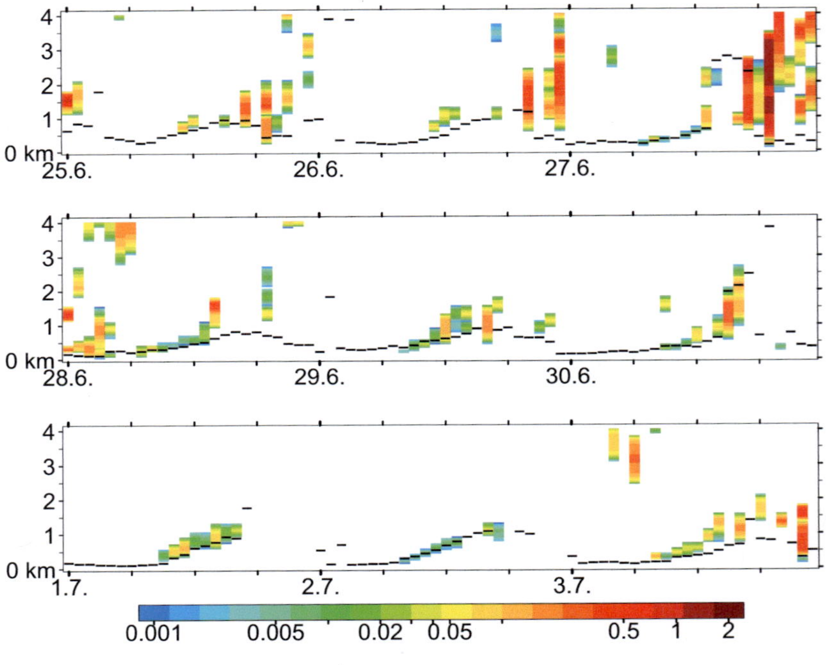

Figure 6.4: Cloud water profile (g kg^{-1}) between 0 and 4 km AGL for the nine-day spin-up time (25 June - 3 July 2016) at Savè. Black bars denote the lowest ceilometer cloud base detection in the corresponding hour.

The ceilometer observations indicate a regular development of the NLLS with a similar diurnal evolution. The modeled elevation of the clouds, in particular the NLLS, matches very well with the ceilometer observations. The results of D2 show daily stratus, although the onset of the cloud formation is lagging behind the observations. In most cases, the modeled clouds do not form before 6 UTC, whereas the observed clouds are fully developed around 3 UTC on average in the time period between 25 June and 3 July (compare Fig. 5.4).
Based on the meteorological insight in D2 and the comparison with D1 we draw the conclusion that D2 is an appropriate basis for the realization of case study simulations. In the following, aerosol feedback simulations are nested into D2 using the setup of D3 (Tab. 3.7).

6.1.2 Aerosol Feedback Modeling Domain (D3)

The D3 setup includes the combination of the TMMS with the prognostic aerosol of COSMO-ART to consider AIE and ADE in the model. Table 6.1 summarizes the D3 realizations that include a variation of the aerosol in ADE and AIE via the factors F_{ADE} and F_{AIE} introduced in Section 3.1.4.

Table 6.1: Summary of the D3 realizations.

Abbreviation	Description
$AIE_{1.0}ADE_{1.0}$	Simulation with $F_{AIE} = 1.0$ and $F_{ADE} = 1.0$ (Reference case).
$AIE_{0.1}ADE_{0.1}$	Simulation with $F_{AIE} = 0.1$ and $F_{ADE} = 0.1$ (Clean case).
$AIE_{1.0}ADE_{0.1}$	Simulation with $F_{AIE} = 1.0$ and $F_{ADE} = 0.1$.
$AIE_{0.1}ADE_{1.0}$	Simulation with $F_{AIE} = 0.1$ and $F_{ADE} = 1.0$.
$AIE_{1.5}ADE_{1.5}$	Simulation with $F_{AIE} = 1.5$ and $F_{ADE} = 1.5$ (Polluted case).

These realizations will be referred to as $AIE_{F_{AIE}}ADE_{F_{ADE}}$ in the following. The realization $AIE_{1.0}ADE_{1.0}$ is taken as a reference since it represents the standard aerosol case. In addition, the two realizations $AIE_{0.1}ADE_{0.1}$ and $AIE_{1.5}ADE_{1.5}$ denote the clean and polluted case, respectively. The terms *clean* and *polluted* have to be seen relatively as part of this experimental setup and imply no general evaluation of the SWA aerosol conditions. The three cases clean, reference and polluted denote physically consistent experiments since AIE and ADE are changed by the same factor ($F_{AIE} = F_{ADE}$). The two simulations $AIE_{1.0}ADE_{0.1}$ and $AIE_{0.1}ADE_{1.0}$ are physically inconsistent ($F_{AIE} \neq F_{ADE}$) but necessary in the framework of the factorial method (Montgomery, 2005) to separate the contributions of AIE and ADE as well as their synergistic effects on the changes in SWA meteorology. The factorial method is introduced in Appendix E together with the equations. Apart from the factorial method, these two simulations were not taken into consideration in the analysis.

6.2 Meteorological and Air Pollution Conditions During 3 July 2016

Regarding D3 the focus is set on 3 July 2016. Nevertheless, also the 2 July 2016 is taken into consideration, since the processes determining the meteorological conditions on 3 July already evolve from the previous day. Therefore the focal time period consists of 31 h from 2 July 15 UTC to 3 July 21 UTC. This time period is spanned by the beginning of the inland propagation of the 2 July EMFE front at 15 UTC, that is relevant for the meteorological conditions in the first half of 3 July, and the arrival of the 3 July EMFE front at Savè supersite that can be taken as the starting point of the NLLS development of 4 July 2016. These 31 h capture therefore the complete time period, relevant for the specification of the atmospheric conditions on 3 July 2016.
Figure 6.5 shows the spatial distribution of total cloud water for the reference case regarding the time period 2 July 18 UTC to 3 July 21 UTC.

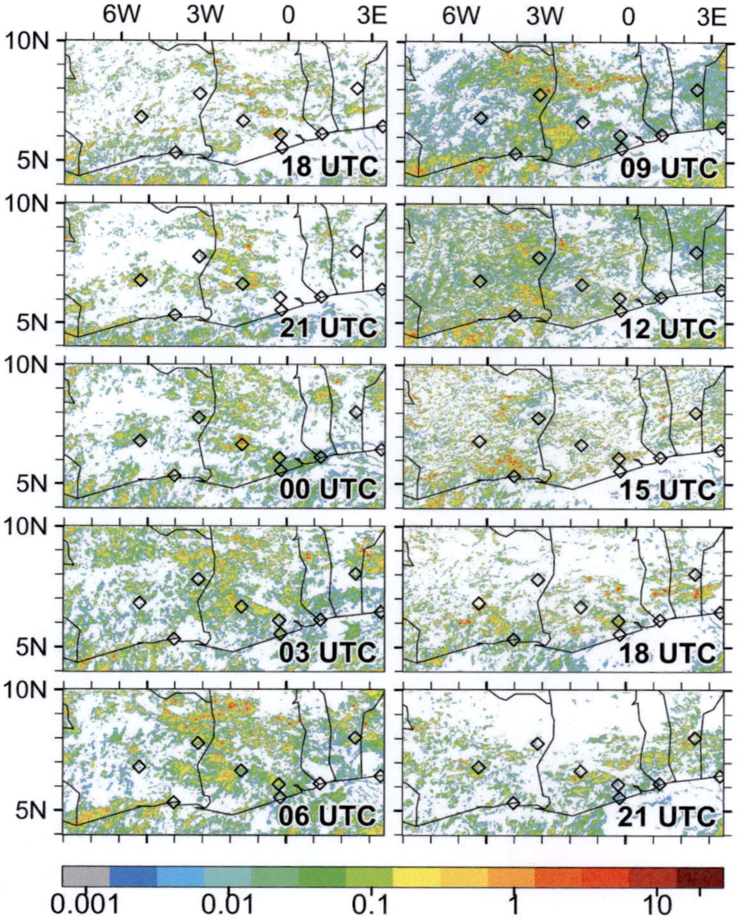

Figure 6.5: Spatial distribution of total cloud water (kg kg^{-1}) for the reference case regarding the time period 2 July 18 UTC to 3 July 21 UTC.

This is meant as a first overview to allow for the identification of features that are subsequently analyzed in more detail, in particular regarding their aerosol sensitivity. During daytime the cloud pattern is characterized by cumuliform clouds. High coverage can be seen, especially in the mountainous areas (Mampong Range, Atakora Mountains and Oshogbo Hills). On 2 July 21 UTC over Ivory Coast a distinct separation between cloudy areas in the South and virtually cloud-free areas in the North are visible. The mechanism related to this feature was introduced as Evening Monsoon Flow Enhancement (EMFE) in Section 2.2 that is linked with an EMFE front and EMFE airmass. Within this study it turned out that the EMFE front is one of the largest meteorological features in SWA, affecting the meteorological conditions over large areas including the NLLS and being also susceptible to changes in the aerosol amount. Section 6.3 shows the findings with regard to the EMFE front.

In the course of the night, the patchy cloud structure transforms to continuous cloud layers indicating stratiform clouds. They develop over large areas of SWA, especially over Ivory Coast, western Ghana, northern Togo and Benin. Between 0 and 3 UTC on 3 July, a coast-parallel onshore cloud layer develops east of 1°E that penetrates inland. In addition to the downstream cloud development in the coastal area, an upstream cloud formation takes place northeast of Savè in the area of the Oshogbo Hills. Beween 10 and 11 UTC, SWA shows highest cloud coverage confirming the findings of Figure 5.4 that indicate a persistent NLLS, lasting till the early afternoon. Ivory Coast is almost completely covered by clouds. Between 12 and 15 UTC the transition from stratiform to cumuliform clouds is visible.

To further quantify the spatiotemporal distribution of SWA clouds, Figure 6.6 presents the SWA land area fraction that is covered by clouds (Fig. 6.6a) and the spatial mean and standard deviation of the cloud base (Fig. 6.6b) for the 31 h time period. A diurnal cycle in the area covered by clouds can be observed (Fig. 6.6a) with a minimum in the late afternoon and early night (around 40 % of the area is covered with clouds) and a maximum in the morning (70-80 % of the area is covered with clouds). Only small differences are visible between the consideration of all clouds (black line) and the clouds below 2 km AGL (green line). This emphasizes the dominance of low-level clouds in SWA in the considered time period.

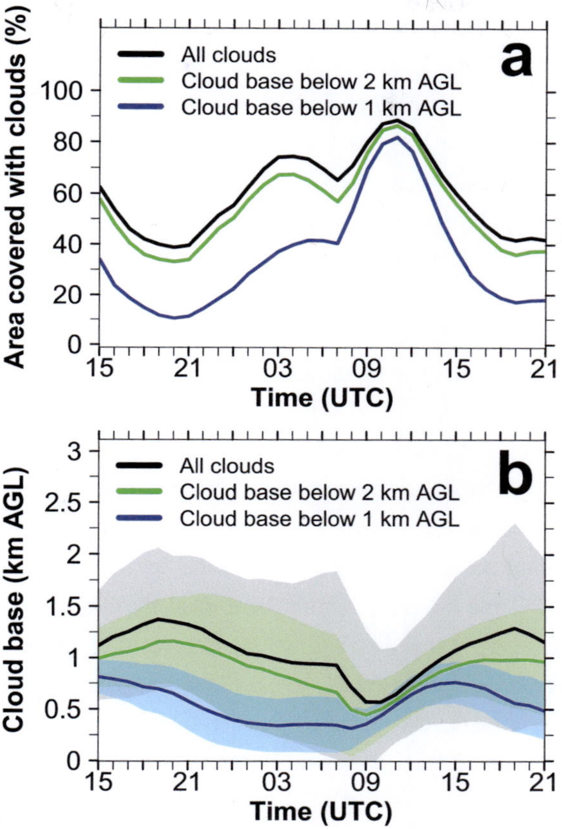

Figure 6.6: Temporal evolution (2 July 15 UTC - 3 July 21 UTC) of (a) the area covered by clouds (%) and (b) the spatially averaged cloud base (km AGL), considering all clouds (black line), clouds with a cloud base below 2 km AGL (green line) and with a cloud base below 1 km AGL (blue line) within the reference case. The analysis in both panels is restricted to the SWA land area (compare Fig. 6.5) and the color shading in (b) denotes the standard deviation.

For the clouds with a base below 1 km AGL, that are taken as a proxy for the NLLS, a similar diurnal cycle is visible but the strongest contribution of these clouds to the total cloud area is restricted to the morning. Between 15 UTC and

21 UTC their contribution is below 20 % but after 21 the contribution gradually increases. After 3 July 7 UTC a steep increase occurs and between 8 UTC and 11 UTC the NLLS dominates the SWA cloud pattern. When comparing with Figure 6.6, it can be seen that the NLLS peak after 7 UTC is related to the development of NLLS over wide areas of Ivory Coast that is delayed compared to the NLLS development in the Savè area. Between 11 UTC and 15 UTC the NLLS contribution to the total coverage significantly decreases, reaching again values below 20 %. The analysis of the cloud base (Fig. 6.6b) shows values around 1.5 km AGL during daytime, typical for fair-weather cumuli indicated by the patchy cloud structure in the corresponding panels of Figure 6.5. The gray shading in Fig. 6.6b indicates a relatively high variation in the cloud base. When focusing on the clouds with a cloud base below 1 km AGL (blue line) the lowest cloud base values are reached between 23 UTC and 8 UTC (approximately 500 m AGL), linked with the gradual increase in cloud coverage in Figure 6.6a. Most interesting is the time period between 8 UTC and 11 UTC. On the one hand the NLLS dominates in the total cloud coverage at this time and on the other hand this time period denotes the smallest standard deviation in the cloud base height. Therefore nearly the entire SWA domain that is covered by clouds, shows a cloud base height around 750 m AGL. After 11 UTC the cloud base increases again with the transition from stratiform to cumuliform clouds. The assessment of the NLLS conditions is realized in Section 6.4.

Figure 6.7 shows the meridional vertical transect along 5°W temporally averaged over the 31 h. The color shading denotes the zonal wind speed component. A distinct transition from the monsoon layer with a westerly component (red shading) to the easterly component (blue shading) above the monsoon layer is visible. The black isolines show the meridional wind component. By considering both components, the southwesterly monsoon flow in the lowest approximately 1.9 km is visible. This is in agreement with the findings at Savè supersite (Kalthoff et al., 2017). Above the monsoon layer, with the dominance of easterly winds, the SAL is located.

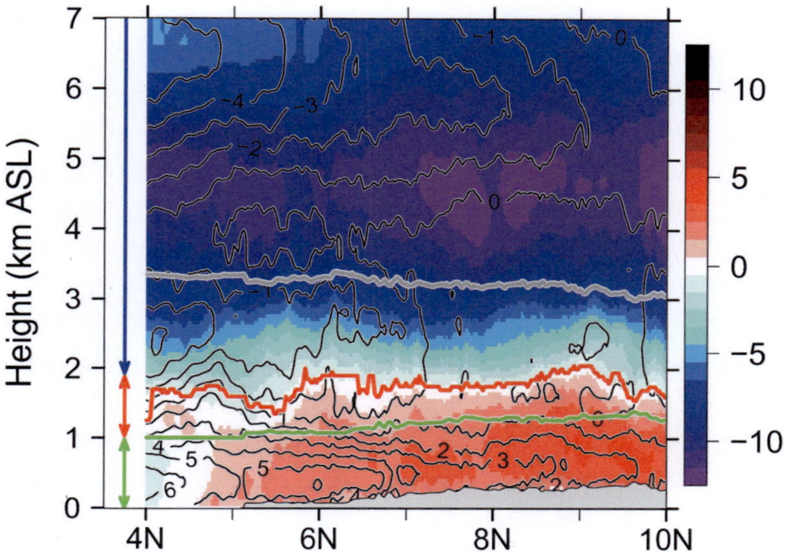

Figure 6.7: Meridional vertical transect (km ASL) along 5°W for the reference case, illustrating the monsoon conditions over SWA averaged from 2 July 15 UTC to 3 July 21 UTC. The color shading denotes the zonal wind speed component (m s^{-1}) and black contours show meridional wind speed component (m s^{-1}). The horizontal lines indicate the height of 1 km AGL as the layer principally affected by the NLLJ (green) and the height of minimum wind speed together with the zonal wind component below zero (easterly wind component) as an indicator for the monsoon layer depth following Fink et al. (2017) (red). Parker et al. (2005a) defined the transition between the monsoon layer and the SAL via the 313 K isoline of virtual potential temperature (gray). The vertical arrows on left hand side denote the three layers used in the following for the vertical discretization of the aerosol: 0-1 km AGL (green arrow), 1-1.9 km AGL (red arrow) and above 1.9 km AGL (blue arrow). The gray shading denotes the topography.

Fink et al. (2017) denotes this transition of the zonal wind component together with a minimum in the wind speed absolute value as the height of the monsoon layer depth (red line in Fig. 6.7). The monsoon layer depth definition of Parker et al. (2005a), using the 313 K isoline of virtual potential temperature, is shown as gray line located around 3.2 km height. The latter rather indicates the beginning of the level at which the maximum of the easterlies is located. In the following, we use the monsoon layer depth definition of Fink et al. (2017). The green line marks the height of 1 km AGL capturing the layer that is primarily affected by the NLLJ as visible from the black contours.

In the following, an introductory overview of the spatial aerosol distribution is presented by using the AOD. With that, the questions about the type of aerosol, their amount, location and possible origin are addressed. Regarding the distribution in vertical dimension, three layers are distinguished: (1) the lowest 1 km AGL that is affected by the NLLJ (green arrow in Fig. 6.7), (2) the layer between 1 km AGL and the top of the monsoon layer at 1.9 km AGL (red arrow in Fig. 6.7) and (3) above the monsoon layer (blue arrow in Fig. 6.7). The combination of (1) and (2) therefore gives information about the total monsoon layer and (3) about the mid and upper troposphere.

Figure 6.8 shows the total AOD as the sum of the contributions from mineral dust, sea salt and anthropogenic aerosol. The spatial AOD distribution reveals a bisection of the domain over land with highest values in the eastern part and lowest values in the western part, especially over western Ivory Coast.

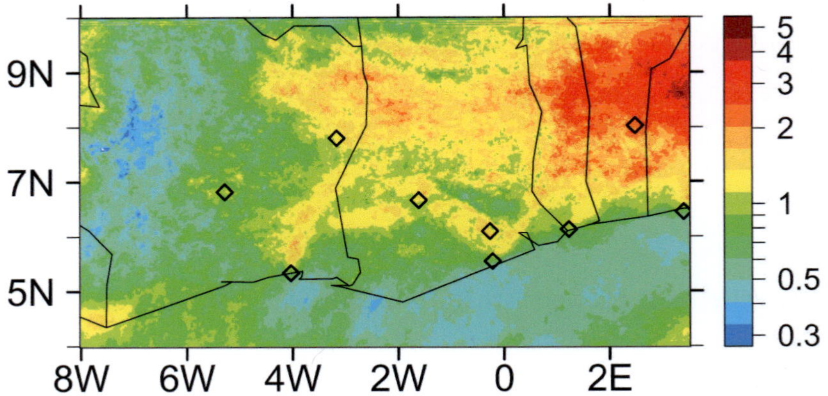

Figure 6.8: Total AOD considering mineral dust, sea salt and anthropogenic aerosol, temporally averaged from 2 July 15 UTC to 3 July 21 UTC for the reference case. The black diamonds indicate the location of selected SWA cities (compare Fig. 2.1).

To shed light on the individual contributions of mineral dust, sea salt and anthropogenic aerosol to the total AOD, Figures 6.9-6.11 focus on the separate aerosol species. The influence of mineral dust in the considered time period is very small (see Fig. 6.9a). From the temporal evolution (not shown) it can be seen, that a tongue of mineral dust enters the domain on 2 July after 18 UTC and propagates westward, covering northern Togo and Benin on 3 July 6 UTC and reaching maximum extent at the end of the considered time period.

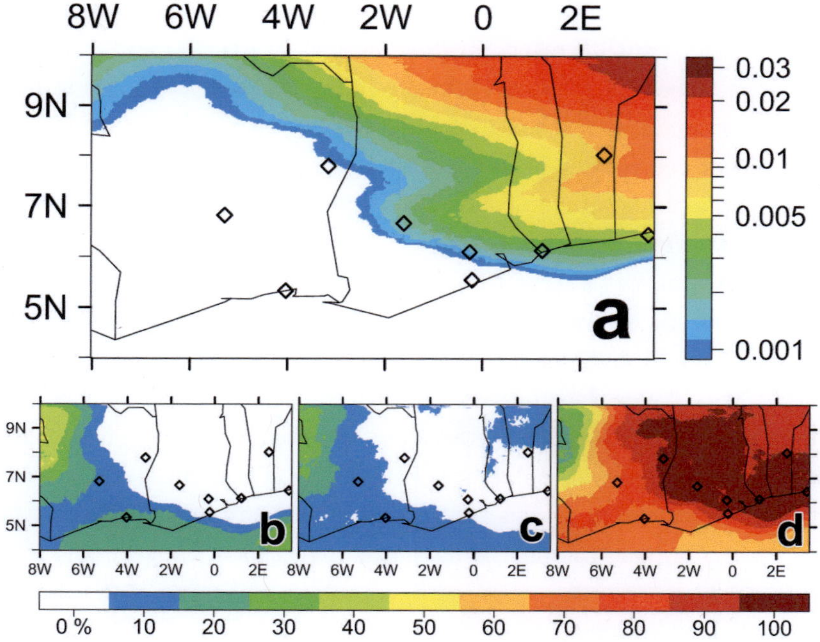

Figure 6.9: Mineral dust AOD (a) temporally averaged from 2 July 15 UTC to 3 July 21 UTC, (b) contribution of the mineral dust AOD in the lowest 1 km AGL to the total mineral dust AOD (%), (c) contribution of the mineral dust AOD in the layer between 1 km and 1.9 km AGL to the total mineral dust AOD (%) and (d) contribution of the mineral dust AOD in the layer above 1.9 km AGL to the total mineral dust AOD (%) for the reference case.

Ivory Coast and southwestern Ghana are virtually unaffected by mineral dust. This behavior is typical for the conditions observed in SWA during the DACCIWA campaign. Mineral dust has generally a low impact on SWA that is primarily restricted to the northern part of the domain. During 3 July 2016 elevated mineral dust concentrations (AOD around 0.2) can be found in a zonal band around 18°N that transports mineral dust to the Atlantic Ocean, as visible from the COSMO-ART D1 realizations (not shown). Figures 6.9b-d denote the contributions to the mineral dust AOD from the layer below 1 km AGL (6.9b), the layer between 1-1.9 km AGL (6.9c) and the layer above 1.9 km AGL (6.9d). This vertical separation emphasizes that mineral dust is relevant for the mid and upper troposphere and has not entered the monsoon layer e.g. via vertical mixing or deposition. Although, the mineral dust does not contribute to the atmospheric composition of the monsoon layer, it can affect the layer via the ADE. Due to the small contribution of mineral dust, this aerosol component is not further investigated in the following.

The AOD from sea salt (Fig. 6.10) is one order of magnitude larger than for mineral dust. As expected, a meridional gradient of the sea salt AOD can be observed with highest values over the Gulf of Guinea and decreasing values inland. Sea salt propagates further inland in the western part of the domain than in the eastern. This is on the one hand due to a stronger southerly component of the near-surface wind speed over Ivory Coast in that time period and on the other hand due to topographic features which are nearer to the coast in the eastern part and can impede the inland propagation (compare Fig. 2.2). These findings of a stronger inland flow in the western part of the domain are confimed by Kalthoff et al. (2017) indicating a stronger NLLJ at Kumasi than at Savè. From the vertical separation it can be seen that in contrast to mineral dust, sea salt clearly dominates in the lowest 1 km AGL (Fig. 6.10b). This highlights the importance of the low-level winds, especially the NLLJ, for the inland transport of sea salt. The temporal evolution of the sea salt AOD (not shown) reveals a clear increase in sea salt burden with the passage of the EMFE front.

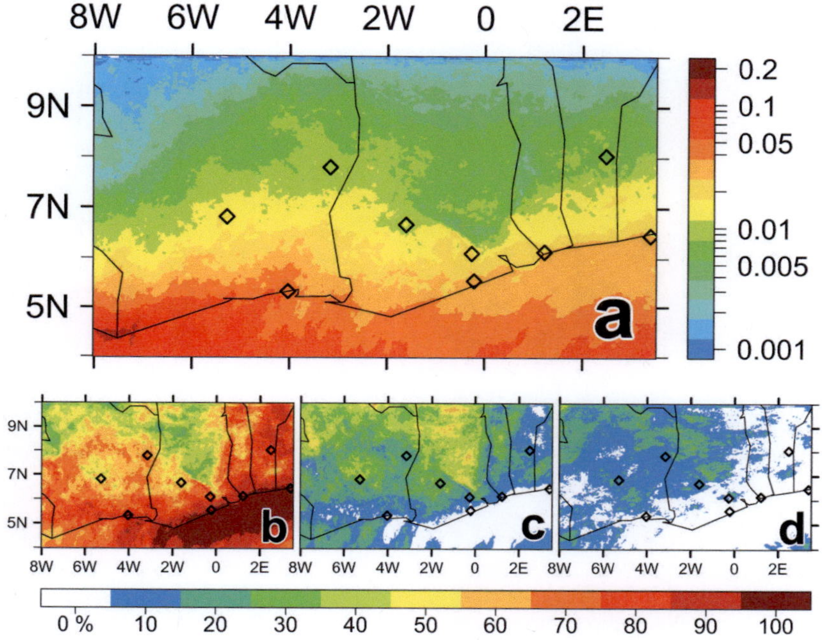

Figure 6.10: Same as Figure 6.9 but for the sea salt AOD. Consider the different scale.

For the eastern part of the domain, sea salt is exclusively located in the lowest 1 km AGL. The topographic features over Ghana (Mampong Range and Atakora Mountains) depict a clear boundary for the inland propagation of sea salt. Therefore north of the Ghanaian mountains, higher amounts of sea salt can be found in the upper level of the monsoon layer (Fig. 6.10c) due to lifting processes at the Mampong Range. Figure 6.11 considers anthropogenic aerosol as the third AOD component. The AOD from anthropogenic aerosol is again one order of magnitude larger than sea salt and therefore clearly dominates the total aerosol picture. Again a gradient is visible with highest values in the eastern part and lowest values in the western part of the domain.

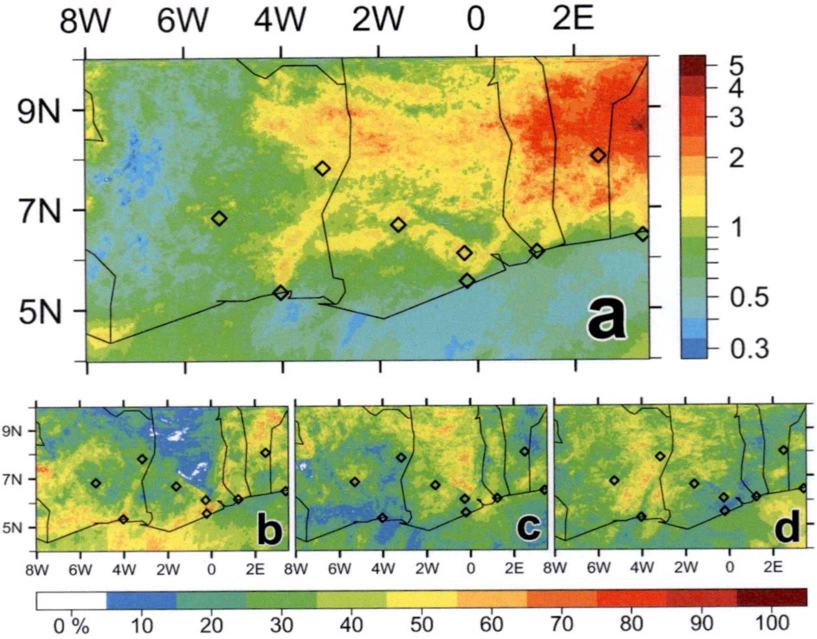

Figure 6.11: Same as Figure 6.9 but for the AOD from anthropogenic origin. Consider the different scale.

Clear city plume signals in northeastward direction can be observed for the coastal cities Abidjan, Accra and Lomé. For Abidjan the plume is detectable over more than 220 km. Furthermore, it seems that higher aerosol concentrations are located upstream of the Mampong Range due to an accumulation upstream of the topographic barrier (Fig. 6.11a). Therefore the near-surface air downstream of the Mampong Range has a significantly lower aerosol burden than the upper monsoon layer (compare Fig. 6.11b and Fig. 6.11c). With the inland propagation of the EMFE front, the post-frontal area is marked by a significant decrease in AOD. The EMFE front pushes the polluted air northwards and replaces it with rather clean maritime air (not shown). In contrast to mineral dust (sea salt), that is most relevant in the upper (lower) levels, the anthropogenic aerosol contributes to all layers (Fig. 6.11b-d). This is linked to a further unique feature of the anthropogenic aerosol: the multiplicity of origins. Sea salt and mineral dust have primarily one source area each, but anthropogenic aerosol is added to SWA via several sources. Firstly, it is emitted locally (e.g. from the cities). Therefore these emissions are injected in the lowest layer, clearly visible in Figure 6.11b. Secondly, it is transported from Nigeria east of the domain via the easterly winds at mid levels and thirdly from the Gulf of Guinea originating from the biomass burning areas in Angola and Dem. Rep. of the Congo. These biomass burning pollution is transported with the lower-tropospheric southeasterly winds and reaches the domain as northward propagating zonal tongues of BC and also CO, visible in the COSMO-ART forecasts (not shown). The biomass burning aerosol influence seems to be more significant in the western part of the domain with their stronger southerly wind component. From the Figures 6.9-6.11 it can be concluded that the analyzed time period is characterized by a bisection of the domain with higher aerosol burden in the east and lower aerosol burden in the west. This bisection is of relevance for the following interpretation of the AIE and ADE effects on SWA meteorology. Mineral dust amounts are located in higher altitudes but show negligible amounts, sea salt is dominating in the lowest 1 km AGL and anthropogenic aerosol is the predominant aerosol component affecting all layers.

The AOD for sea salt and especially anthropogenic aerosol are surprisingly high. The following excursus is added to shed light on the causes of these elevated AOD values.

6.2.1 Excursus: Causes of Elevated AOD Values Modeled by COSMO-ART

The AOD is analyzed in more detail by focusing on 3 July 6 UTC as an example for a time with elevated AOD values. The total AOD for the entire vertical column (Fig. 6.12a) shows values of 5.8 in the 99th percentile. When comparing with the AOD contribution of the monsoon layer (lowest 1.9 km AGL, Fig. 6.12b) it can be seen that a large contribution is related to this layer. Furthermore, Figure 6.12c masks out the areas of Fig. 6.12a that are covered by clouds. Figure 6.12c reveals that the areas with highest AOD values, emphasized by the gray isoline denoting an AOD of 2, are primarily related to cloudy areas. Therefore these AOD values can neither be captured by observations from sun photometers nor from satellite.

In clouds or areas with very high RH the AOD have not necessarily to be on the order of magnitude that is expected by observations from sun photometer or remote sensing since in moist regimes, water uptake of aerosol occurs. The increase in aerosol liquid water content (ALWC) leads to an increase in aerosol size (hygroscopic growth) that alters the AOD and therefore the ADE (e.g. Bian et al., 2014).

Further insight in the aerosol properties is realized in Figure 6.13 via the analysis of vertical profiles of wet aerosol mass density, aerosol number density and wet and dry aerosol median diameter at two locations in SWA denoted via the black circles in Figure 6.12. Location L1 is related to an area with an elevated AOD (5.9) and location L2 exhibits a smaller AOD (0.7). As shown in Figure 6.12c, both locations are masked by clouds. In Figure 6.13 the two locations are represented by the red (L1) and green (L2) curves. The analysis is separated in the different COSMO-ART aerosol modes whereas the coarse modes SEASB, SEASC, SOILB, and SOILC are neglected due to their small contribution.

The mass densities (Fig. 6.13a-g) show a sharp and significant increase around 1.5 km AGL related to the monsoon layer.

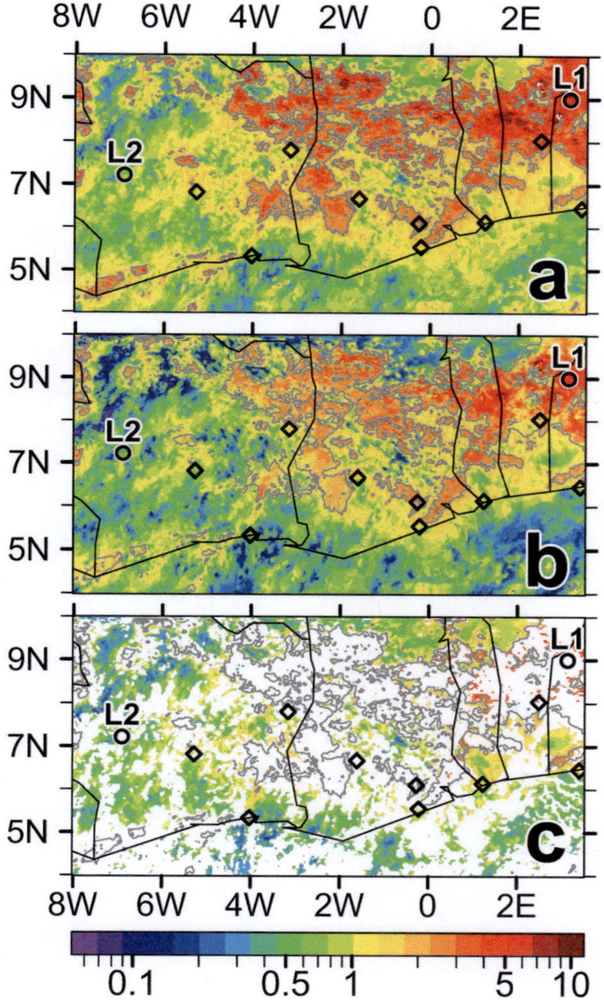

Figure 6.12: Total AOD of the reference case on 3 July 6 UTC for (a) the entire vertical column, (b) the lowest 1.9 km AGL (monsoon layer) and (c) as for (a) but by masking the areas covered by clouds. Gray contours denote areas with AOD values above 2. The circles L1 and L2 show locations used for analyzing vertical profiles of aerosol properties (Fig. 6.13).

ACC and ACCS show largest contributions to the total mass shown in Figure 6.13g. The contribution from AIT and AITS is one order of magnitude smaller and the contribution of the coarse modes SEASA and SOILA is below 10 μg m^{-3}. For L2, located in the western part of SWA, no mineral dust signal is visible (Fig. 6.13f and 6.13g). The total mass density (Fig. 6.13g) shows for the entire vertical profile lower values for L2 but especially in the monsoon layer large differences occur which comprise more than two orders of magnitude. In contrast to the significant differences in the mass density, the number density is similar for L1 and L2. However, for the lowest 3 km AGL L1 shows lower values for AIT, AITS and ACCS as L2. Near the surface, remarkable high number densities around 10.000 cm^{-3} are reached that are primarily related to new particle formation in the aitken mode.
The aerosol median diameter is given in Figure 6.13o-t with and without considering the ALWC (solid and dashed line, respectively). Again the strongest differences occur in the monsoon layer. In this layer the wet aerosol median diameters are about one order of magnitude larger than the dry median diameters. In contrast, the dry aerosol median diameters show minor differences between L1 and L2. Since mineral dust is hydrophobic, COSMO-ART does not consider the water uptake of mineral dust (Fig. 6.13t). From a spatial analysis for 6 UTC it is obvious that the water uptake in the accumulation mode, realized by ISORROPIA II in COSMO-ART, shows the strongest increase when the RH reaches and exceeds 95 % (not shown). This is also visible when comparing the dry and wet median diameters in ACC and ACCS (Fig. 6.13q,r) with the profile of the relative humidity (Fig. 6.13u). A sharp increase in the wet aerosol median diameter occurs below 2 km AGL when the RH reaches 95 %.

Figure 6.13 ***(facing page)*****:** Vertical profiles (km AGL) at L1 (green, compare Fig. 6.12) and L2 (red, compare Fig. 6.12) of (a-f) wet aerosol mass density (μg m^{-3}), (h-m) aerosol number density (cm^{-3}) and (o-t) aerosol median diameter (μm) for dry aerosol (dashed line) and wet aerosol (solid line) considering the aerosol modes AIT, AITS, ACC, ACCS, SEASA, SEASB and SEASC on 3 July 6 UTC for the reference case. The total mass and number density are shown in (g) and (n), respectively. Additionally, the RH profile (%) is given in (u).

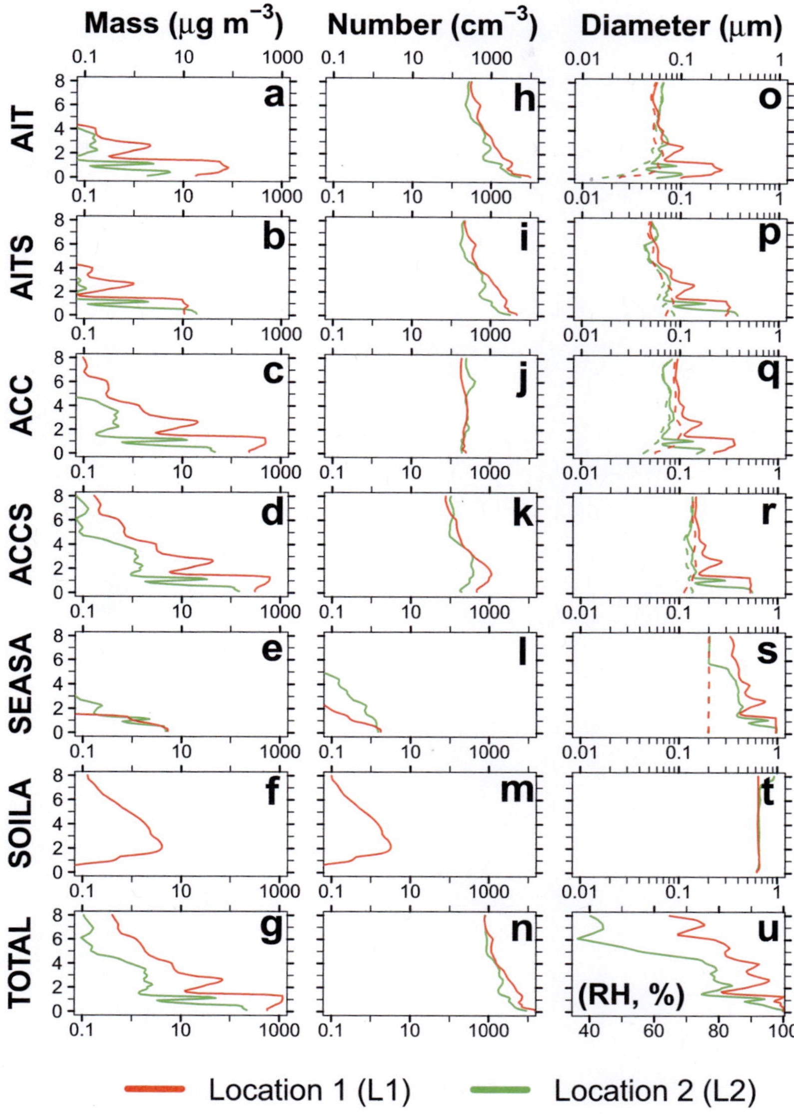

Mass (μg m^{-3})
Number (cm^{-3})
Diameter (μm)
AIT
AITS
ACC
ACCS
SEASA
SOILA
TOTAL
a
b
c
d
e
f
g
h
i
j
k
l
m
n
o
p
q
r
s
t
u
(RH, %)
Location 1 (L1)
Location 2 (L2)

From the overview of the mass density, number density and median diameter profiles it can be concluded that the monsoon layer marks the strongest differences between L1 and L2 with a higher mass and number density in L1 compared to L2. With the overall higher RH at L1, the wet aerosol median diameters at L1 are generally higher which means a larger ALWC contributes to the total mass density.

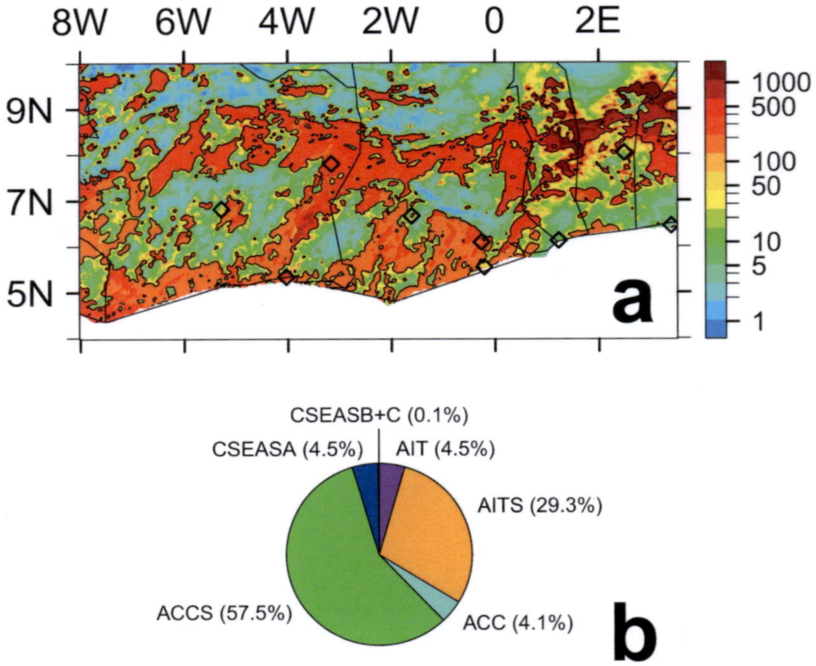

Figure 6.14: Analysis of the ALWC at 500 m AGL over land for the reference case on 3 July 6 UTC. (a) Total ALWC (μg m^{-3}, shading) and RH of 95% (black contour) and (b) pie chart of the ALWC contribution from the single aerosol modes (%) to the total ALWC in (a).

Although, the near-surface mass and number densities are high, they are not unrealistic and also the wet aerosol median diameters does not reach magnitudes at which they would rather relate to cloud droplets than to aerosol particles (approximately 10 μm). In the last part of this excursus the spatial distribution of ALWC is analyzed. Figure 6.14a shows the spatial distribution of total ALWC at 500 m AGL, a height that is related to high differences between dry and wet aerosol median diameters (Fig. 6.13o-s). Since the focus is related to the characteristics over land, the ALWC over the ocean is masked. The black contour highlights areas with a RH above 95 %. It is obvious that highest ALWC values can be found in areas with a RH exceeding 95%. Furthermore, the areas with the highest ALWC mainly coincide with the areas of the highest total column AOD (Fig. 6.12). The 99th percentile of ALWC is 0.88 mg m^{-3}. Seinfeld and Pandis (2006a) indicated that the concentration of water in the aerosol phase is less than 1 mg m^{-3}. Bian et al. (2014) focused on a case study of ALWC in the North China Plain, an area that is likewise affected by high air pollution, by using ISORROPIA II that is also used in COSMO-ART. They found average ALWC values of (0.169$\pm$0.164) mg m^{-3} and the maximum value of 0.971 mg m^{-3}. These literature values are in agreement with the findings of COSMO-ART. Bian et al. (2014) also indicates that the ALWC values, measured from aircrafts in the North China Plain, can even be higher, reaching median values of 27 mg m^{-3} in stratocumulus clouds (Deng et al., 2009). Figure 6.14b shows the ALWC contribution of the single aerosol modes to the total ALWC presented in Figure 6.14a. The pie chart reveals that the aged particles in the aitken and accumulation mode (AITS and ACCS) clearly dominate the ALWC contribution (86.8 %). Only minor contributions come from AIT, ACC and sea salt.

The highest values of AOD can be found in the eastern part of the domain (Fig. 6.11). This area is on the one hand enveloped by the Atakora Mountains and the Oshogbo Hills and on the other hand the three large cities Accra, Lomé and Abidjan (a megacity) emit their pollution plumes into this area, clearly visible in Figure 6.11a,b. That means a high aerosol amount is emitted in an area with impeded air exchange due to topographic features. Furthermore, the aerosol is released in an anvironment with high RH leading to an enhanced water uptake on aerosols that is even further increased with the lifting at topographic barriers and the corresponding increase in RH at the Oshogbo Hills, located perpendicular to the monsoon flow. The diurnal cycle of the total AOD (not shown) reveals an AOD maximum during night further underlining the relevance of the ALWC to the total AOD. During night, the PBL height decreases, leading to a higher aerosol burden in the near-surface layer, a layer that is simultaneous related to an increase in RH due to radiative and advective cooling. Also the western part of the domain is enveloped by mountains, in particular the Mampong Range that is also perpendicular to the monsoon flow. But the area framed by the mountains is larger and only one big city, Abidjan, emits their pollution plume into this area. Therefore the aerosol emission is nonuniform distributed in the research area with a clear surplus in the eastern area.

The analysis of the aerosol properties in this excursus, considering the mass and number density, the wet and dry aerosol median diameter and the ALWC, does not reveal indications for erroneous AOD estimations in COSMO-ART. It rather emphasizes the need to distinguish between the AOD that is observed in cloud-free conditions from sun photometer and satellite and the modeled AOD that also captures cloud-affected areas. Furthermore, it underlines the importance of the ALWC contribution to the aerosol diameter and therefore the total aerosol mass density, especially when focusing on the very moist monsoon layer in SWA. The spatially irregular distribution of aerosols, with higher burden in the eastern part and lower burden in the western part provide a possibility to study the cloud characteristics under different aerosol conditions with at the same time similar meteorological conditions.

Additional insight in the atmospheric pollution situation on 3 July is provided by the analysis of pollution from gas flaring. Based on the gas flaring parameterization, that was developed within this study (Sect. 3.1.3), Figure 6.15 shows the results of a sensitivity study for the D2 domain to assess the relevance of gas flaring in the overall atmospheric composition. The ozone contribution from gas flaring (Fig. 6.15a) is up to 35% in the Niger Delta area, reaching values of about 15 ppbv. NO_x (Fig. 6.15b) shows very high contributions and affects large areas along the African coast. The gas flaring hot spot is the Niger Delta but also in the offshore area of Ivory Coast individual flaring sites are located. Although the gas flaring parameterization does not include direct aerosol emissions, new particles can be formed in COSMO-ART. Figure 6.15c shows the gas flaring contribution due to particles in the mode AIT for the Niger Delta subdomain as emphasized by the black rectangle in Figure 6.15b including the gas flaring sites as black circles. The results indicate that the contribution from gas flaring to the total atmospheric composition is non-negligible, especially in the coastal regions. Figure 6.15a-b furthermore reveals that the gas flaring pollutants from the Niger Delta are lifted to greater heights and can be transported westwards with the easterly winds, reaching the southern areas of Ivory Coast and Ghana.

6.3 Aerosol Affecting the Evening Monsoon Flow Enhancement (EMFE)

In the following, the characteristics of EMFE and their interaction with changes in the aerosol is quantified. Section 6.3.1 assesses the aerosol-dependent inland penetration of the EMFE front. The potential feedback mechanism between EMFE and the aerosol is described in Section 6.3.2. In the last section of this part the aerosol impact on EMFE-related clouds and precipitation is analyzed (Sect. 6.3.3).

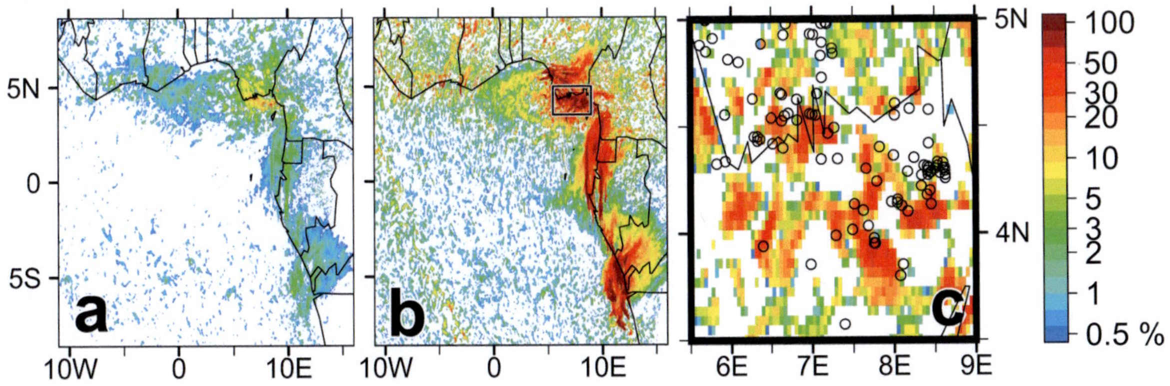

Figure 6.15: Contribution (%) of pollution from gas flaring to the total air pollution (vertically integrated) on 3 July 12 UTC after a 36 h D2 simulation with and without flaring consideration for (a) O_3, (b) NO_x and (c) aerosol in mode AIT for the Niger Delta (black rectangle in (b)). The location of gas flaring sites is indicated by black circles.

6.3.1 Location and Propagation of the EMFE Front

Following Grams et al. (2010), the EMFE front position can be estimated by the location at which a specific isentrope of virtual potential temperature $\theta_{v,s}$ crosses a specific height h_s. For this study the potential temperature θ_s=302 K and the height h_s=250 m are used.

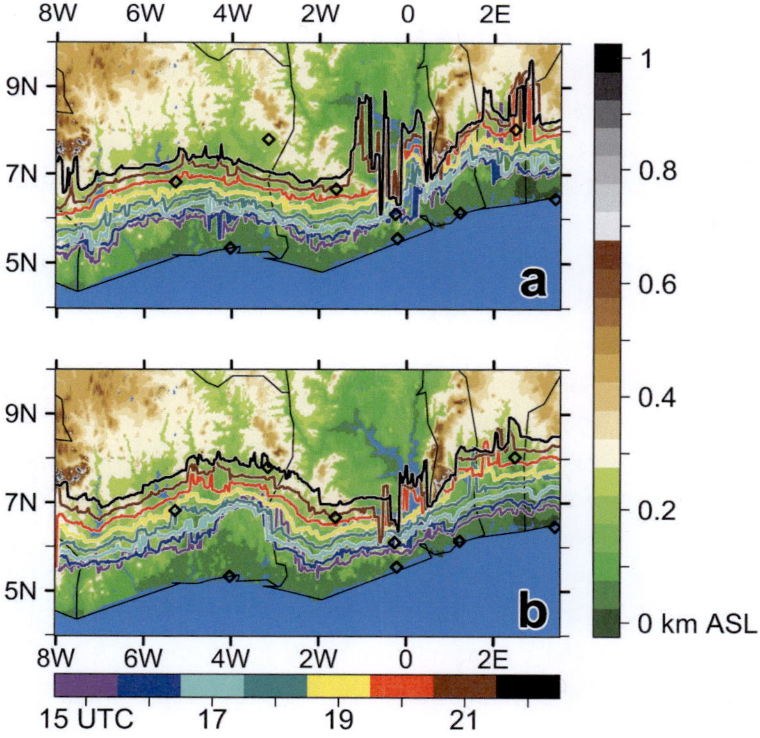

Figure 6.16: Localization of the EMFE front at (a) 2 July 2016 and (b) 3 July 2016 between 15-22 UTC for the reference case. The front is detected by the arrival of the isentropic surface θ_s=302 K at h_s=250 m AGL. The color of the front denotes the time (UTC, bottom table of colors). The underlying shading shows the topography of SWA (km ASL, right table of colors).

These values are related to the COSMO-ART results of this study and might be different for other modeling studies. Figure 6.16 shows the location of the EMFE front between 15 and 22 UTC for 2 July 2016 (Fig. 6.16a) and 3 July 2016 (Fig. 6.16b). The results for 3 July are added to underline that the EMFE is a robust feature occurring frequently over SWA, which is also indicated by the study of Adler et al. (2017). Nevertheless, the main focus is related to the EMFE front on 2 July since this front has the possibility to affect the NLLS developing in the night from 2 to 3 July 2016 that is of interest in this study.
The θ_s method for the EMFE front location is only an approximation since the potential temperature is also altered by the surface characteristics and diabatic effects. Therefore we focus on the time period 15-22 UTC that coincides well with the wind speed anomalies as shown subsequently in Figure 6.21. With the increasing nighttime cooling over land after 22 UTC, the temperature gradient between the EMFE postfrontal and prefrontal airmass diminishes, impeding the localization of the front. Figure 6.16 reveals that the frontal propagation derived from the θ_s method is more homogeneous over Ivory Coast and western Ghana than in the eastern part of the domain. Especially over Lake Volta, with its relatively low temperatures but also reduced roughness, the method shows higher variability. At 15 UTC the EMFE front is located about 100 km inland. After 15 UTC, the front propagates inland nearly coast parallel but is affected by land surface features (e.g. Atakora Mountains and Lake Volta).
To quantify the differences in the frontal characteristics between the three COSMO-ART realizations (clean, reference and polluted case), Figure 6.17 shows the frontal velocity averaged over Ivory coast (7.5-3.0°W) for 2 July (Fig. 6.17a) and 3 July (Fig. 6.17b). In general an increasing velocity between 16 and 19 UTC can be seen (accelerated motion phase). After sunset (18:32 UTC at Abidjan on 2 July 2016) the EMFE front receives no further acceleration and the velocity stays constant (uniform motion phase). On 2 July (3 July) the EMFE front velocity stagnates around 7 m s^{-1} (8 m s^{-1}).

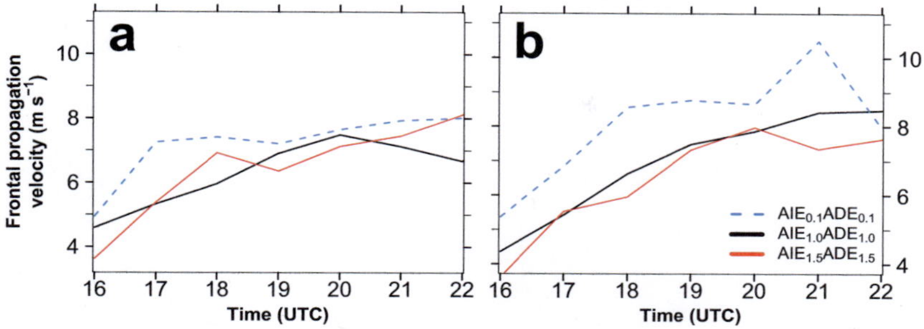

Figure 6.17: Temporal evolution of the EMFE frontal velocity (m s^{-1}) for (a) 2 July and (b) 3 July 15-22 UTC, based on the frontal detection of Figure 6.16 and zonally averaged over Ivory Coast (7.5-3.0°W) for the clean (blue dashed), reference (black solid) and polluted case (red solid).

This is on the same order of magnitude when comparing with the findings of Grams et al. (2010) of 10 ± 1 m s^{-1} for the frontal propagation in the Mauritanian area. The three realizations show a similar temporal evolution. Remarkable is the time lag in the transition from the accelerated motion to the uniform motion that is already reached at 17 UTC for clean case (black dashed line in Fig. 6.17a), at 18 UTC for the polluted case (red line) and 20 UTC for the reference case. On 3 July, the three realizations show more distinct different velocities and a clearer pattern of reduced velocity with increasing aerosol. On both days the clean case shows highest velocities and the aerosol scaling factor increase from 0.1 to 1 has a significant stronger impact on the EMFE frontal velocity than the increase from 1 to 1.5. It has to be considered that the increase from 0.1 to 1.0 is related to a factor of 10 and the increase from 1.0 to 1.5 to factor 1.5. Therefore it cannot be concluded whether the relatively small difference between the reference and the polluted case is related to an aerosol saturation or to the unequal intervals of the factors. Further model realizations are necessary in this respect. The results imply a stronger but shorter acceleration phase for the clean case and a longer and more moderate acceleration phase for the experiments with higher aerosol amounts. Figure

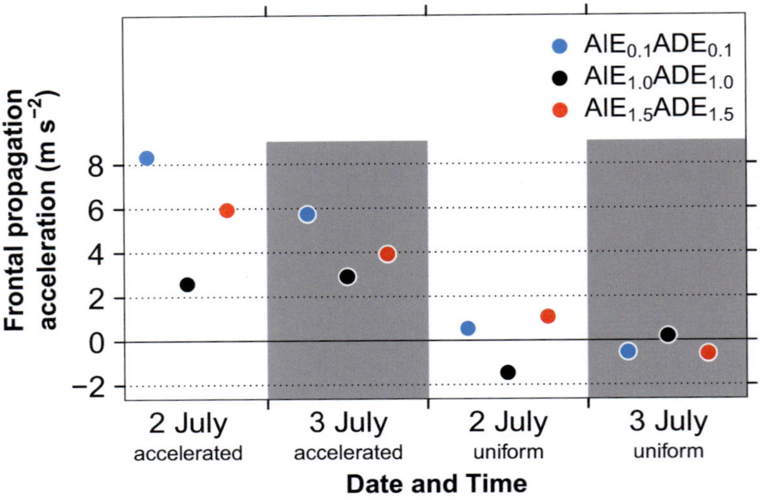

Figure 6.18: Temporal evolution of the acceleration of the EMFE front (m s^{-2}) for 2 July 2016 (white background) and 3 July 2016 (gray background) averaged over Ivory Coast (7.5-3.0°W). The dots denote the average acceleration for the clean (blue), reference (black) and polluted case (red) regarding their individual accelerated and uniform motion phase.

6.18 underlines the findings of the velocity analysis in Figure 6.17. While the accelerated motion phase is related with positive accelerations around 3-8 m s^{-2}, the uniform motion phase shows generally negligible accelerations. The polluted case is not confirming the general trend. From Figure 6.17 it can be seen that the variability in the diurnal evolution is relatively high, raising the need to conduct a larger number of realizations to come to more robust conclusions. To assess the question, whether the spatial mean of the EMFE front velocities are statistically significant different between the realizations, a Wilcoxon-Mann-Whitney rank sum test with a significance level of α=0.05 was applied for 2 July, considering the spatial variations of the EMFE fronts over Ivory Coast. Figure 6.19 summarizes the results in which green boxes indicate a statistically significant difference between the velocity averages of a

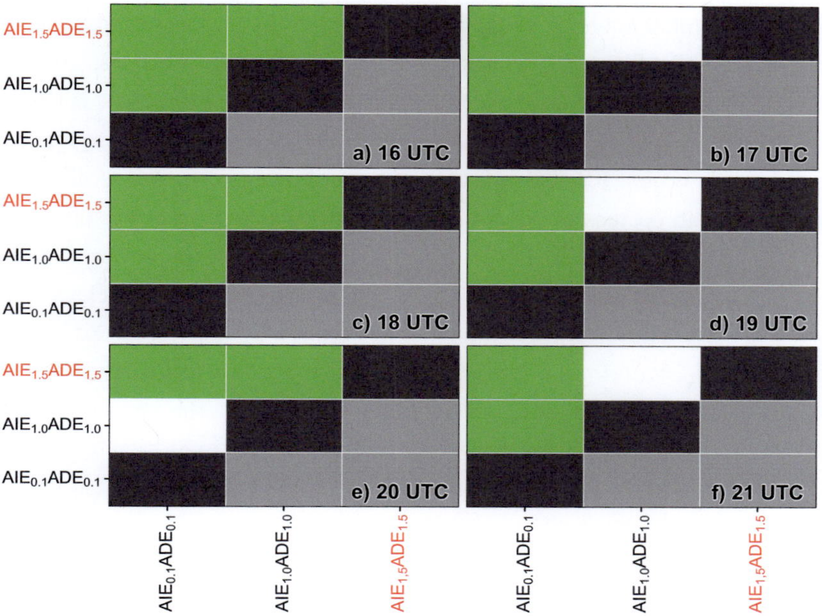

Figure 6.19: Statistical significance of the pairwise difference between the experiment averages of the EMFE front velocity for 2 July 16-21 UTC. Green boxes indicate a significant difference whereas white boxes denote no significant difference (Wilcoxon-Mann-Whitney rank sum test with α=0.05). The velocity at 16 UTC is calculated from the frontal location difference between 15 and 16 UTC (analog for the other times).

pair of experiments. The figure reveals that the differences between the clean case and the cases with higher aerosol are all statistical significant. Between the reference and the polluted case the differences are partly statistically insignificant due to the large variations. For completeness, the significance test is repeated for the acceleration of the EMFE front. The results are added in Appendix D. For the acceleration, statistical significant differences are identified especially in the first hours (strongest acceleration) (Fig. D.1). The distance between the coast and the inland propagating EMFE front is shown in Figure 6.20 for all realizations. Between 15-16 UTC the location of the EMFE fronts are similar but after 16 UTC a divergence can be observed. In the early afternoon the EMFE front is located stationary inland near the coast. Around 15 UTC it starts to propagate inland. At this time the coastal distance is lowest for the clean case and increases with more aerosol. After 16 UTC a significant acceleration occurs for all realizations. As described above, the acceleration increases with decreasing aerosol and therefore the EMFE front in the clean case (in the beginning behind of the other EMFE fronts) is now ahead. At 22 UTC a spatial shift of about 25 km established between the clean case and the cases with higher aerosol. By considering a frontal velocity of around 7 m s^{-1}, the EMFE front of the reference and polluted case reach the location of the clean case EMFE front with a delay of one hour. The impact of these frontal shifts on EMFE-related precipitation is described in Section 6.3.3.

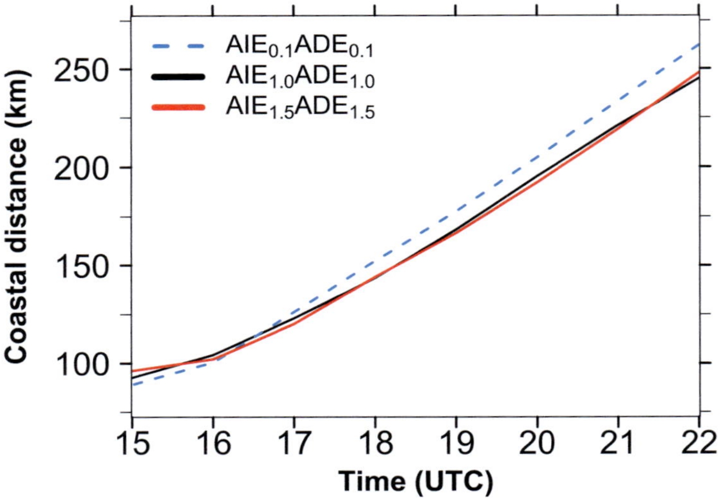

Figure 6.20: Temporal evolution of the inland propagation of the EMFE front via the distance from the coast (km) on 2 July 2016 between 15-22 UTC, spatially averaged over Ivory Coast (7.5-3.0°W) and by considering the clean (blue dashed), reference (black solid) and polluted case (red solid).

6.3.2 Assessment of the Aerosol-EMFE Feedback Mechanism

First Indication

After identifying changes in the EMFE frontal location with changing aerosol amounts, the question about the underlying feedback mechanism arises. The aerosol impact on the EMFE front becomes obvious when analyzing the near-surface wind speed differences. The anomaly between the clean and the reference case shows a filament structure in zonal direction that propagates northward with time and captures nearly the entire SWA domain. Figure 6.21a shows the 250 m AGL wind speed anomaly (clean minus reference case) on 2 July 22 UTC together with the location of the EMFE fronts.
When concentrating on the western part of the domain, a coherent pattern can be observed with a spatial shift between the two EMFE fronts. As already indicated by Figure 6.20, the clean case EMFE front is ahead of that of the reference case. Furthermore, the postfrontal area is characterized by higher wind speeds that the prefrontal area. Therefore the area between the fronts shows a distinct band of positive wind speed anomalies up to 9 m s^{-1}. The contribution from ADE and AIE as well as the synergistic effects (SYN) of ADE and AIE to the wind speed differences oberved in Figure 6.21a are assessed via the factorial method (Montgomery, 2005). The description of this method is given in Appendix E. Figure 6.21b-d present the factorial method results in terms of the percentaged contribution from AIE, ADE and SYN, respectively. The factorial method reveals that the anomalies are predominantly caused by ADE. This can clearly be seen in the area between the two EMFE fronts. Only minor contributions come from AIE and SYN. The contributions in the remaining domain are patchy and randomly distributed (not shown). To allow for clearness, only the areas with wind speed differences above 1 m s^{-1} are shown in Figure 6.21b-d. The factorial method therefore provides a first hint on the potential mechanisms between EMFE and aerosol, indicating that the interaction between aerosols and radiation are more important than the

interaction between aerosols and clouds. These strong responses on changing aerosol have not been expected beforehand and the hypothesis is raised that this strong response is dominating all subsequent aerosol-induced changes in SWA, in particular the characteristics of the NLLS. In Figure 6.21a the eastern part of the domain do not show these coherent patterns. This is likely related to the topographic barriers and Lake Volta. East of 1°E also an anomaly band of wind speed is visible but less pronounced than in the western part.

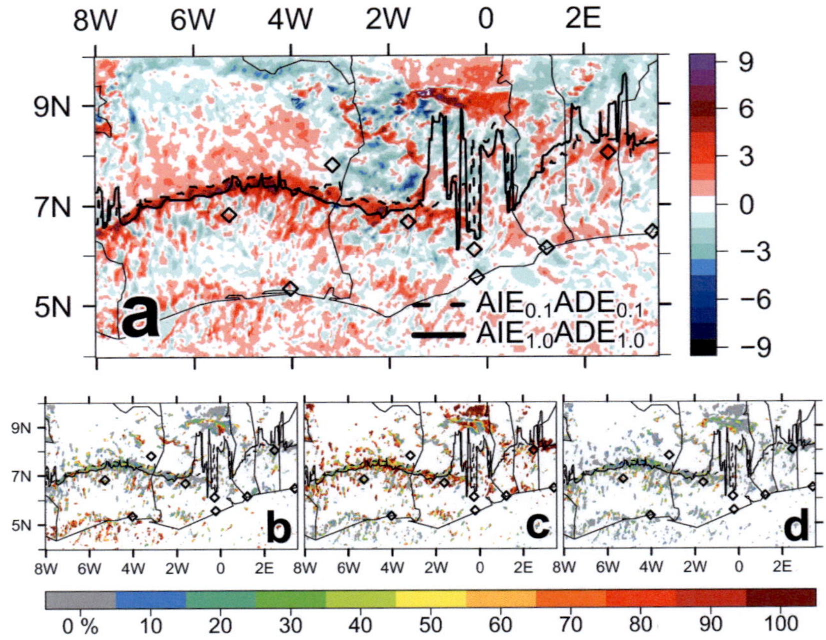

Figure 6.21: Analysis of the wind speed at 250 m AGL on 2 July 22 UTC including (a) the wind speed difference (m s^{-1}) between clean and reference case, (b) the contribution of AIE (%) to the changes in (a), (b) the contribution of ADE (%) to the changes in (a) and (d) the contribution of SYN (%) to the changes in (a). Panels (b)-(d) are derived from the factorial method by focusing on areas with a wind speed difference greater than 1 m s^{-1} in (a). The black dashed (solid) line shows the EMFE front for the clean (reference) case.

Evidence for the Dominating ADE

The stationarity of the EMFE front near the coast in the early afternoon is related to the balance between the landward directed monsoon flow over the ocean and the turbulence over land (e.g. Grams et al., 2010). Therefore a change in the turbulence can alter the balance and lead to differences in the EMFE front propagation (turbulence mechanism). Since the ADE is dominating the identified changes, it can be expected that the surplus in surface heating within the clean case due to a reduced extinction of downward shortwave radiation leads to stronger land-sea pressure gradients that could affect the EMFE front propagation (pressure gradient mechanism).

To shed light on these questions, Figure 6.22 shows the differences in surface net downward shortwave radiation (Fig. 6.22a), 2 m temperature (Fig. 6.22c) and sea level pressure (Fig. 6.22e) between the clean and the reference case. In addition, the Figures 6.22b,d,f show the surface net downward shortwave radiation, 2 m temperature and sea level pressure, respectively, only for the areas that are cloud free in both realizations to exclude effects from displaced clouds and to highlight the ADE in a cloud-free environment. The differences in the surface meteorological quantities reveal a clear signal. With decreasing aerosol, less downward shortwave radiation is scattered and absorbed, leading to an average increase in surface net downward shortwave radiation of about 69 W m^{-2} but can reach significant higher values locally (Fig. 6.22a-b). In the eastern part of the domain, that is related to higher AODs (Fig. 6.8) and is therefore more sensitive to aerosol changes, the increase in radiation is even more pronounced. The increase in incoming shortwave radiation leads to an increase in the 2 m temperature up to 6.6 K with a mean of +0.6 K (Fig. 6.22c-d). The temperature increase furthermore leads to a domain wide reduction in surface pressure up to 1.1 hPa with a mean of -0.23 hPa over land (Fig. 6.22e-f). To prove the hypothesis that the surface pressure difference is caused by the temperature difference, the barometric formular together with the pressure and temperature at 850 hPa were applied to calculate the surface

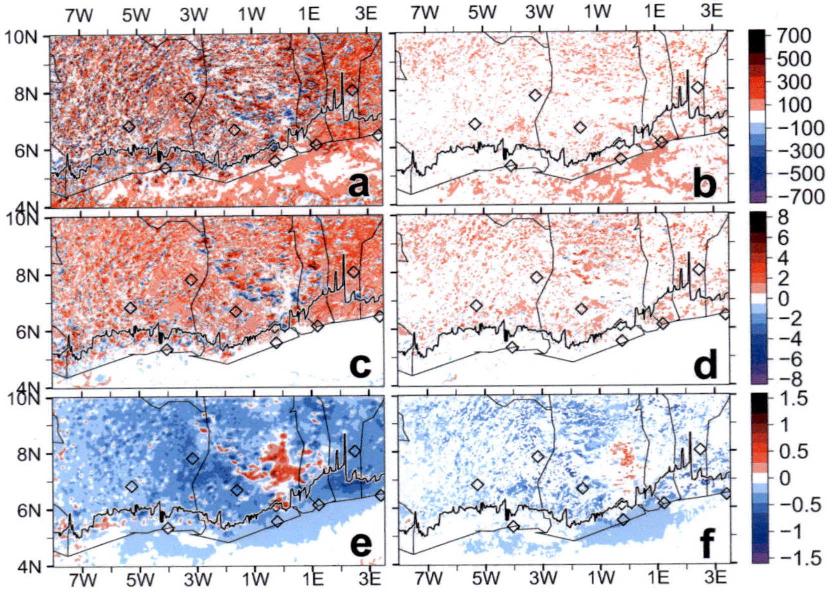

Figure 6.22: Surface meteorological quantities on 2 July 15 UTC as differences between the clean and the reference case ($AIE_{0.1}ADE_{0.1}$ - $AIE_{1.0}ADE_{1.0}$) over SWA including cloudy and cloud-free areas (left) and over the areas that are simultaneously cloud free in the clean and reference case (right). (a-b) Surface net downward shortwave radiation difference (W m^{-2}), (c-d) 2 m temperature difference (K) and (e-f) sea level pressure difference (hPa). The black solid line denotes the location of the reference case EMFE front.

pressure over Ivory Coast at 15 UTC (compare Fig. 6.22f). The hydrostatic approach provides the spatially averaged value -0.21 hPa that is in agreement with the modeled value of -0.23 hPa. It can be concluded that the pressure changes are dominated by the changes in temperature. The negative pressure anomaly is less pronounced in southwestern Ivory Coast due to less amounts of aerosols. North of the Mampong Range and the Atakora Mountains, especially over Lake Volta, a positive pressure anomaly establishes. This is related to the fixed SSTs, suppressing a near-surface temperature change (Fig. 6.22c). With the development of a local heat low in the surrounding areas, over Lake Volta a local high develops. Due to the fixed SSTs also over the Gulf of Guinea no temperature changes are visible.

Detailed Insights in the EMFE Frontal Properties

A more detailed insight in the EMFE front characteristics is realized by meridional transects over central Ivory Coast (5.75°W) and central Benin (2.25°E) at 15 UTC (early phase of the EMFE front, Fig. 6.23) and 21 UTC (late phase of the EMFE front, Fig. 6.24). The transects include wind speed, pressure, moisture, radiation and heat fluxes for the clean case (blue solid line) and reference case (red solid line). Furthermore, the EMFE fronts of the clean and reference case are shown as vertical blue and red dashed lines, respectively. The vertical black solid line denotes the location of the coastline, indicating that for the eastern transect the coastlines is located more than 1° farther north. At 15 UTC the EMFE fronts are located near the coast (Fig. 6.23). The differences between the EMFE postfrontal and prefrontal areas are characterized by a marked postfrontal increase in the meridional wind speed component at 250 m AGL and a postfrontal increase in RH but not in specific humidity. At 15 UTC a near-surface temperature gradient has developed between land and sea (with fixed SSTs). A pronounced increase in surface net downward shortwave radiation in the clean case is visible (as indicated by Figure 6.22c-d) that is higher in the eastern part of the domain with the higher aerosol burden. Therefore the increase in surface net downward shortwave radiation increases the sea-land temperature gradient and the sensible heat flux as an indicator for the turbulence. Finally, the difference in the heating leads to a sea level pressure decrease over land. The pressure difference between the clean and reference case is highlighted as green line. The stronger heating in the eastern part of the domain results in a stronger pressure reduction. Six hours later at 21 UTC the EMFE fronts have propagated approximately 1° inland (Fig. 6.24). The EMFE postfrontal area is still characterized by a marked jump in the wind speed at

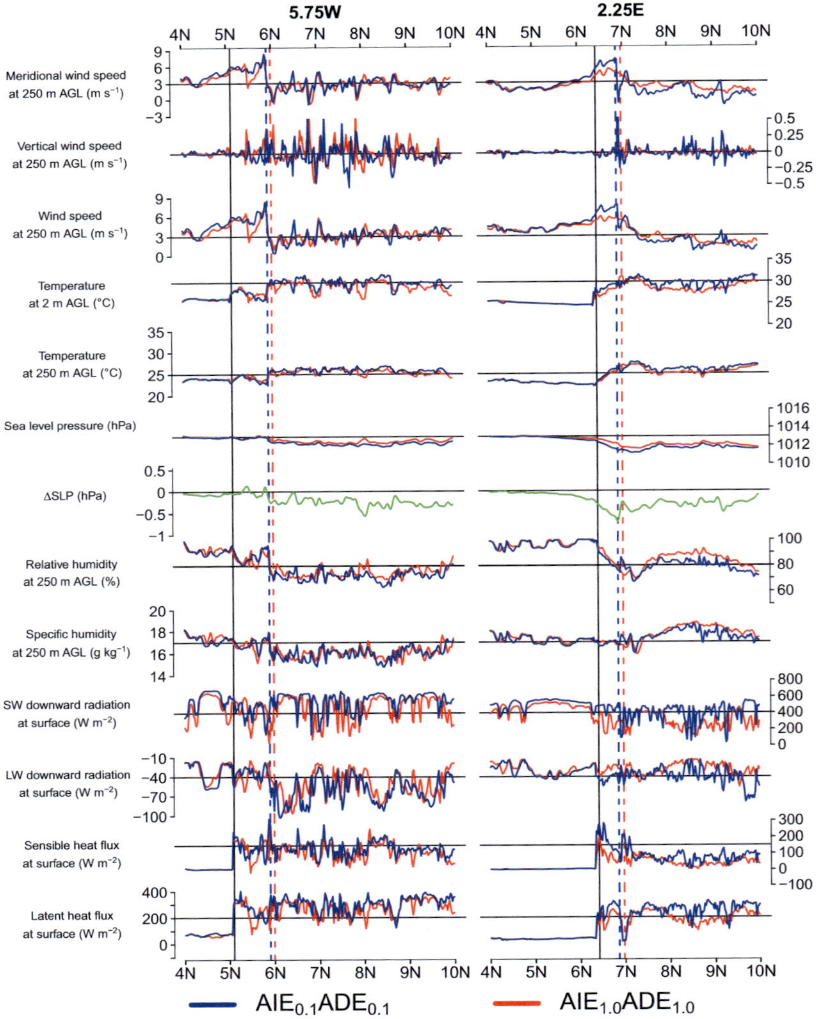

Figure 6.23: Meridional transect (4°N-10°N) of meteorological quantities along 5.75°W (central Ivory Coast, left) and along 2.25°E (central Benin, right) for 2 July 15 UTC and the clean (blue) and reference case (red). The mean sea level pressure (SLP) difference between the clean and the reference case is shown in green. The coastline is indicated as vertical black line and the EMFE front location of the clean and reference case are given as vertical blue and red dashed lines, respectively. Within this study the sensible and latent heat flux are defined positive upward.

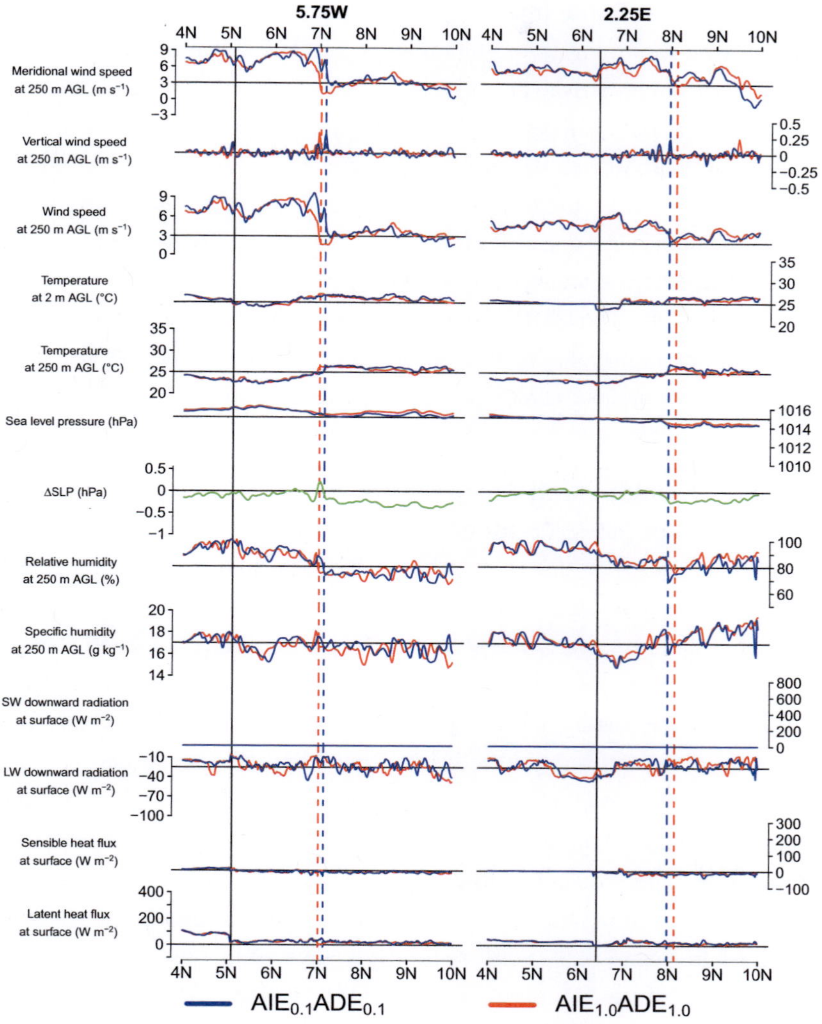

Figure 6.24: Same as Figure 6.23 but for 2 July 21 UTC.

250 m AGL. For the western transect the clean case EMFE front is ahead of the reference case EMFE front, clearly visible in the spatial shift of the frontal peak in wind speed. For the eastern transect no clear signal of the shift in the EMFE front can be derived due to strong zonal variations.
The eastern transect is located close to the Savè supersite and therefore these results can contribute to the question whether the observed increase in RH during night is related to the advection of cold air or air that transports higher specific humidity. The transects clearly show that the postfrontal area is related to higher RH corresponding to reduced temperatures but not to an increase in specific humidity. The EMFE seems to act as a compensator of the aerosol-induced differences that have developed over land during the day, as visible for the temperature and pressure.

The Two Counteracting Components of the Aerosol-EMFE Feedback

In the beginning of this section it was suggested that the two mechanisms *turbulence* and *pressure gradient* can be the cause of aerosol-related changes in the EMFE front location. In fact, the two mechanisms are counteracting. The stronger the heating, the stronger the turbulence in the PBL and therefore the stronger the impact on the EMFE front balance between the monsoon flow and the turbulence over land with respect to a suppression of the inland propagation. Regarding the pressure gradient, a stronger heating increases the sea-land pressure gradient, shifting the EMFE front balance to the opposite direction and favoring the inland propagation. From the meridional transects presented in Figure 6.23, both mechanisms are visible - the increase in the sensible heat flux and the decrease in sea level pressure. When going back to the temporal evolution of the coastal distance of the EMFE front relating to the coherent pattern over Ivory Coast (Fig. 6.20), both mechanisms are visible intrinsically. The effect of the turbulence mechanism can be seen in the early afternoon when the EMFE front is located virtually stationary over land near the coast (15 UTC and earlier). The lower the aerosol amount, the stronger the heating and the stronger the turbulence. This shifts the balance between the monsoon flow and turbulence to the latter, leading to a stationary EMFE front location nearer to the coast. With an increase in aerosol, the reduced shortwave heating shifts the balance to the monsoon flow, leading to a stationary EMFE front farther away from the coast. Unfortunately, the EMFE front detection via the θ_s method fails for the time period earlier than 15 UTC. Therefore the total difference in the stationary EMFE front location with changing aerosol cannot be assessed. Nevertheless, it is an interesting thought that the location of the EMFE front in their stationary phase over Ivory Coast, relative to the coastline, could be used as a proxy for the aerosol burden in that area (under the otherwise same conditions).

For the time period after 16 UTC, when the turbulence decreased sufficiently, the pressure gradient mechanism seems to dominate because the EMFE front in the clean case becomes faster than in the other cases. As visible in Figure 6.17, at 16 UTC the EMFE front in the clean case has already a higher velocity than in the cases with increased aerosol, indicating that the EMFE front in the clean case overcomes the suppression from turbulence faster and starts to propagate inland earlier than in the other cases due to the stronger pressure gradient force. Figure 6.25 summarizes the counteracting components turbulence and pressure difference that govern the inland propagation of the EMFE front by comparing the temporal evolution of the clean and reference case differences in surface sensible heat flux (red, positive downward) and surface pressure (blue), spatially averaged for the EMFE prefrontal area over Ivory Coast. The temporal evolution clearly shows that the sensible heat flux differences (and the absolute values itself) decrease more strongly with time than the pressure differences. After sunset the sensible heat flux is zero but the pressure difference maintains till the EMFE front has passed the area (compare Fig. 6.24).

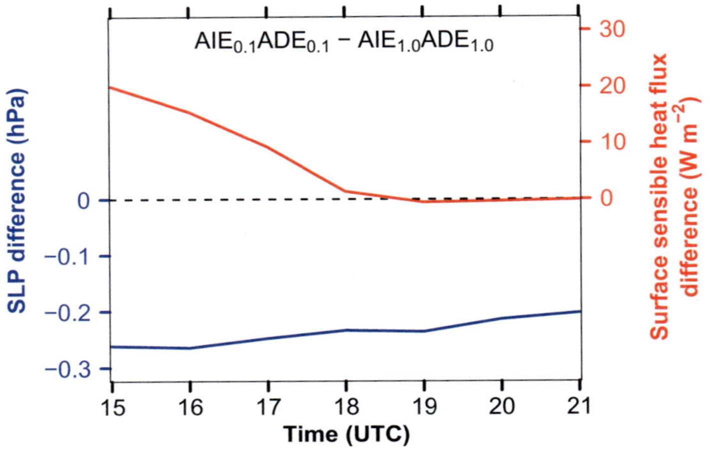

Figure 6.25: Temporal evolution of the differences in surface sensible heat flux (red, W m^{-2}) and surface pressure (blue, hPa) between the clean and reference case for the time period 2 July 15-21 UTC spatially averaged for the EMFE prefrontal area over Ivory Coast. The sensible heat flux is defined positive downward.

The monsoon flow over SWA is driven by the temperature gradient between the cool SSTs of the eastern equatorial Atlantic Ocean and the SHL. The modeling domain D3 is located in between these extremes neither changing the SSTs that are fixed or the SHL that is not part of D3. With this location, D3 therefore can serve as an amplifier that is able to increase or decrease the temperature gradients and thereby the EMFE front characteristics. In agreement, Grant and van den Heever (2014) showed that the sea breeze front over Cameroon weakens with enhanced aerosol number concentration. The study of Adler et al. (2017) revealed that a frontal structure frequently occurs under undisturbed monsoon conditions over SWA, reaching Savè on average around 21 UTC. This agrees well with the latitudinal EMFE front location in the eastern transect (Fig. 6.24, dashed lines on right hand side).

Effects From Longwave Cooling

The mechanisms described above focus on the impact of shortwave radiation during daytime. However, also the longwave radiation after sunset can influence the EMFE characteristics. It is expected that an aerosol reduction leads to enhanced longwave cooling due to less longwave absorption and re-emission from the aerosol layer. A reduction of the near-surface temperature could favor the formation of the stable nocturnal boundary layer and therefore a decoupling between the near-surface layer and the residual layer, further accelerating the EMFE front in the clean case due to less friction.

The surface net downward longwave radiation difference between the clean and reference case does not show coherent patterns as observed for the shortwave radiation (Fig. 6.22a,b). For example on 2 July 15 UTC (not shown) even the cloud-free areas reveal patchy anomaly patterns for the longwave radiation difference. The longwave radiation and temperature difference for three selected times are presented in Figure 6.26 (2 July 19 UTC), Figure 6.27 (2 July 23 UTC) and Figure 6.28 (3 July 5 UTC).

The figures follow the structure of Figure 6.22 in which the left hand side shows the meteorological quantities for cloudy and cloud-free areas whereas the right hand side only shows areas that are simultaneously cloud free in the clean and reference case. The longwave radiation difference is calculated by subtracting the absolute values of the clean and reference case. Positive values (red) therefore indicate more outgoing longwave radiation (OLR) in the clean case and negative values (blue) more OLR in the reference case. The black line shows the EMFE front of the reference case. In the following, the focus is set on Ivory Coast. Figure 6.26 refers to 19 UTC, shortly after sunset. The EMFE prefrontal area shows positive and negative OLR differences (Fig. 6.26a,b) but the red areas (more OLR in the clean case) dominate. The EMFE prefrontal 2 m temperature anomaly shows still higher values for the clean case as observed at 15 UTC (Fig. 6.22c,d), although the anomaly has weakened. The EMFE postfrontal area exhibits virtually no temperature differences,

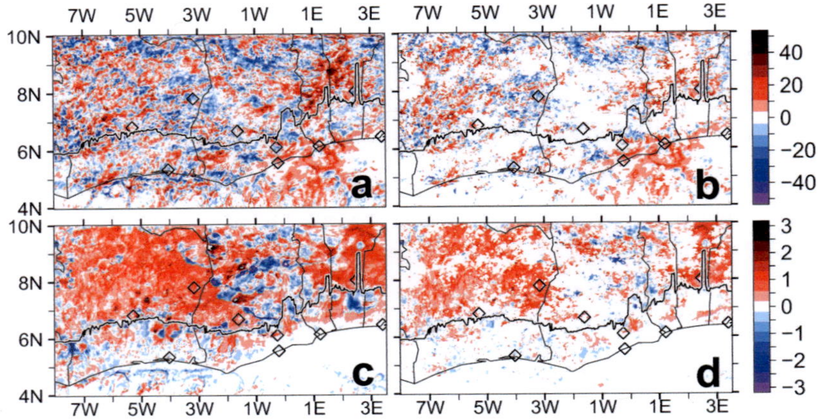

Figure 6.26: Surface meteorological quantities on 2 July 19 UTC as differences between the clean and the reference case ($AIE_{0.1}ADE_{0.1}$ - $AIE_{1.0}ADE_{1.0}$) over SWA including cloudy and cloud-free areas (left) and over the areas that are simultaneously cloud free in the clean and reference case (right). (a-b) Surface net downward longwave radiation difference (W m^{-2}) and (c-d) 2 m temperature difference (K). For the calculation of the longwave radiation difference the absolute values were subtracted. A positive (negative) value therefore relates to more OLR in the clean case (reference case). The black solid line denotes the location of the reference case EMFE front.

confirming the findings identified above that the EMFE compensates the differences that have developed during daytime.

At 23 UTC, approximately 5 hours after sunset, blue areas dominate in the EMFE prefrontal area, indicating more OLR in the reference case (Fig. 6.27a,b). Generally more OLR for the reference case is visible for cloud-free areas and more OLR for the clean case is obvious for cloudy areas. It appears that a positive OLR anomaly for the clean case in the prefrontal area only occurs in the first hours after sunset. Due to the aerosol reduction, the absorption and re-emission from the aerosol is reduced and the energy loss into space occurs faster. Therefore the clean case OLR is smaller than for the reference case later in the night. In cloudy areas, areas characterized by high RH, the aerosol can significantly grow due to the ALWC (see Sect. 6.2.1).

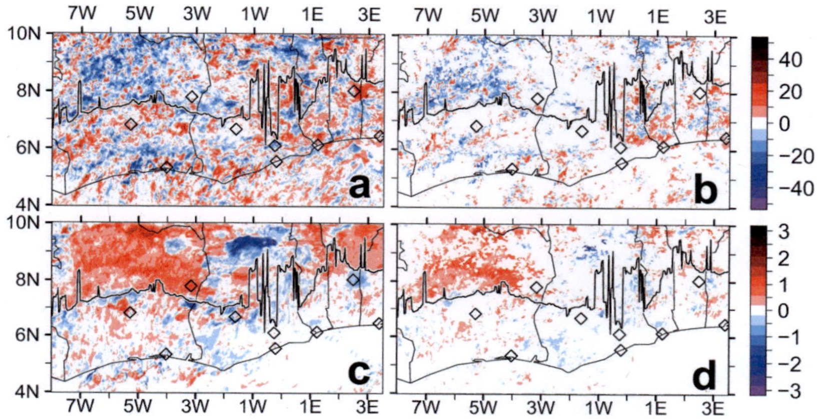

Figure 6.27: Same as Figure 6.26 but for 2 July 23 UTC.

This can substantially affect the radiation budget. Therefore the radiation is more susceptible to changes in the aerosol amount for the cloudy areas (e.g. the EMFE postfrontal area). In the cloudy areas the aerosol reduction leads to a positive OLR anomaly for the clean case since more OLR can be transmitted through the cloud layer than in the reference case. The positive 2 m temperature anomaly in the clean case is still present in the EMFE prefrontal area (Fig. 6.27c,d). On 3 July 5 UTC, approximately 11 hours after sunset and shortly before sunrise, the EMFE front has passed SWA entirely. The positive temperature anomalies in the clean case, resulting from the surplus in shortwave heating during the day, were compensated by the advection of the cold EMFE airmass (Fig. 6.28c,d). The green rectangles in Figure 6.28b,d highlight a cloud-free area with a temperature response in the corresponding area of a change in OLR. A spatial average over this rectangle reveals a positive OLR anomaly of 6.8 W m^{-2} that is linked with a negative temperature anomaly likely due to longwave cooling of -0.17 K.

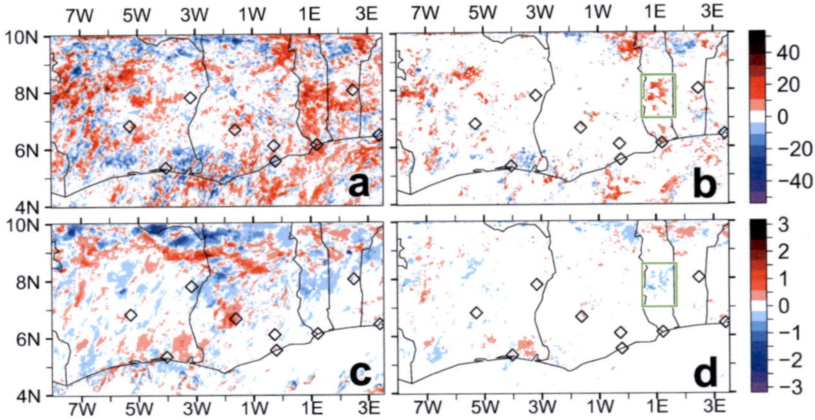

Figure 6.28: Same as Figure 6.26 but for 3 July 5 UTC. The green rectangle emphasizes an area used for a detailed analysis as described in the text.

The results indicate that the higher OLR of the clean case in the first hours after sunset can cool especially the EMFE prefrontal area. However, the strong positive inland temperature anomaly in the clean case, resulting from the shortwave heating during daytime, persists attenuated also during night. The longwave cooling may reduce but cannot compensate this positive anomaly. It persists until the EMFE airmass compensates the differences between the clean and the reference case. Therefore the positive temperature anomaly superpose potential effects from longwave cooling. In the EMFE postfrontal area, in which the positive temperature anomaly is compensated, no intensified large-scale nighttime longwave cooling is visible for the clean case in the COSMO-ART results.

Convergence at the EMFE Front

To analyze the vertical structure of the EMFE front, Figure 6.29 shows the temporal evolution of the convergence from the meridional wind speed component at the EMFE front between 0-1500 m AGL, derived from the 20 km post- and prefrontal area over Ivory Coast on 2 July 15-22 UTC.

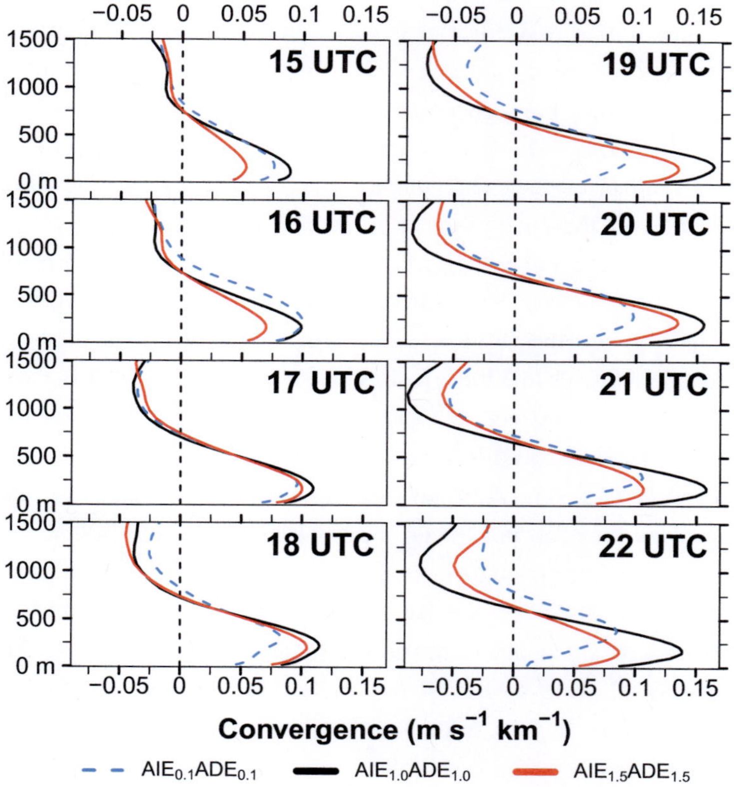

Figure 6.29: Temporal evolution of the convergence profile (m s^{-1} km^{-1}, height in m AGL) at the inland propagating EMFE front using the meridional wind speed component spatially averaged 20 km behind and 20 km ahead of the EMFE front on 2 July 2016 between 15-22 UTC, spatially averaged over Ivory Coast (7.5-3.0°W) and by considering the clean (blue dashed), reference (black solid) and polluted case (red solid). The vertical black dashed line emphasizes a zero convergence.

Convergence generally can be found below 750 m AGL with the maximum around 250 m AGL. The temporal evolution shows weaker convergence in the afternoon that increases in the course of the evening. Grant and van den Heever (2014) shows comparable convergence values of 0.05-0.15 m s^{-1} km^{-1} below 750 m AGL for a study over Cameroon. Regarding the different aerosol scenarios, no distinct pattern is visible. After 16 UTC the reference (clean) case shows the highest (lowest) convergence. The polluted case is located in between. Striking is the strong increase in convergence after sunset (18 UTC) in the reference and polluted case. Figure 6.29 suggests a stronger EMFE front in the cases with more aerosol but it has to be considered that an increase in convergence can be the result of a reduced prefrontal wind speed or an enhanced postfrontal wind speed. Therefore, Figure 6.30a additionally shows the absolute wind speed values prefrontal (dashed) and postfrontal (solid) for the three realizations at 250 m AGL that is related to the maximum convergence in Figure 6.29. As expected, in all realizations the postfrontal wind speed is higher than the prefrontal wind speed. Whereas in the convergence (Fig. 6.29) no clear aerosol trend is visible, the absolute values of the pre- and postfrontal wind speed (Fig. 6.30a) indeed reveal a trend. Generally the pre- and postfrontal wind speeds decrease with increasing aerosol. At 15 UTC, with a rather stationary front, the differences between the realizations are small. In the course of the evening and night, the pre- and postfrontal wind speeds of the realizations diverge. The increase in convergence after 16 UTC in the reference and polluted case (Fig. 6.29) are not related to an intensification of the postfrontal winds exceeding the values of the clean case, but to a weakening of the prefrontal winds, compared to the clean case. This different behavior is especially obvious after sunset (around 18 UTC) with a further decrease in the prefrontal wind speeds (green and red dashed lines in Fig. 6.30a). While the postfrontal winds increase, the prefrontal winds decrease, leading to the stronger convergence. For the clean case the pre- and postfrontal wind speeds have a virtually constant bias and therefore the temporal variation in the convergence is small. To further illustrate the

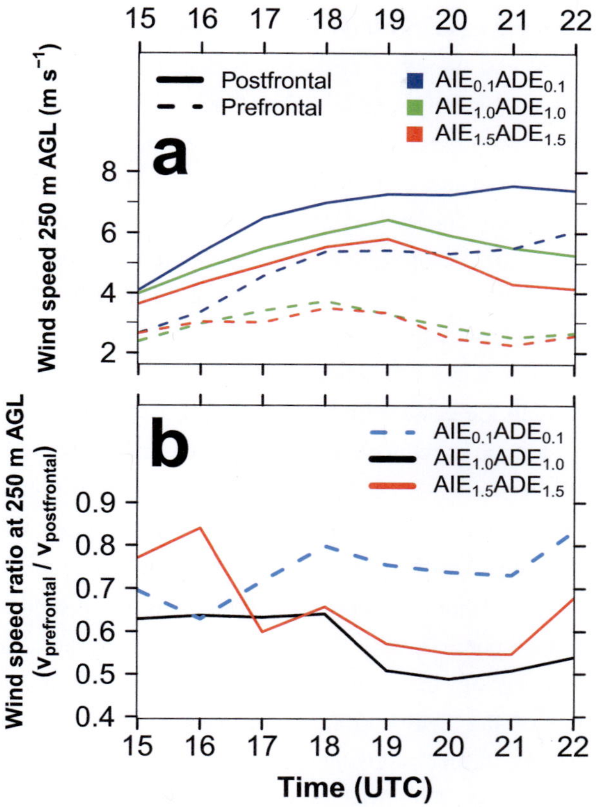

Figure 6.30: Temporal evolution of (a) meridional wind speed (m s^{-1}), 20 km ahead of the EMFE front (prefrontal, dashed) and 20 km behind the EMFE front (postfrontal, solid), spatially averaged over Ivory Coast (7.5-3.0°W) on 2 July 15-22 UTC at 250 m AGL (according to the height of the maximum convergence in Fig. 6.29) for the clean (blue), reference (green) and polluted case (red) and (b) the ratio of prefrontal and postfrontal meridional wind speed considering the clean (blue dashed), reference (black solid) and polluted case (red solid).

results, Figure 6.30b shows the ratio of the prefrontal and postfrontal wind speed at 250 m AGL. The lower the ratio, the stronger the differences in pre- and postfrontal wind speeds and therefore the stronger the convergence. The main difference that can be observed between the reference and polluted case on the one hand and the clean case on the other hand is that the prefrontal winds increase significantly weaker with the increase in postfrontal winds as in the clean case. After sunset even a decrease in pre- and postfrontal winds in the reference and polluted case can be observed whereas in the clean case the wind speeds stay relatively constant. For the interpretation it has to be considered that the EMFE front is propagating and with that the area for the wind speed calculation changes. At 15 UTC it is near the coast and at 22 UTC it is around 7°N linked with an increase in topographic height of around 250 m ASL.

We speculate that these differences are related to a combination of two mechanisms: 1. In the cases with higher aerosol it was shown that the SWA near-surface heating is reduced via a reduction in surface net downward shortwave radiation (Fig. 6.22a-b). Therefore the reduced pressure gradient force leads to weaker EMFE pre- and postfrontal wind in the afternoon. 2. Via the ADE the nighttime radiative cooling is reduced for higher aerosol burden. For the realizations with more aerosol therefore a weaker decoupled near-surface atmospheric layer is expected in contrast to the clean case. The weaker the decoupling, the stronger the surface friction that likely leads to a deceleration of the EMFE front in the cases with higher aerosol. The EMFE front in the clean case likely conserves the momentum obtained in the afternoon via the decoupling from the surface.

For the period 17-19 UTC, capturing the time around sunset with a rather stable propagation of the EMFE front (between the strong acceleration in the afternoon and the weakening in the night; compare Fig. 6.30b), the following linear relationship can be derived between the dimensionless aerosol scaling factor $F = F_{AIE} = F_{ADE}$ and the 20 km postfrontal meridional wind speed $\bar{v}(F)$ averaged over Ivory Coast

$$\bar{v}(F) = -1.11F + 6.59; \; [\bar{v}] = m\, s^{-1}. \tag{6.1}$$

Equation 6.1 is not stated to claim general applicability but to gain insight in the modeled wind speed response to different values of F_{AIE} and F_{ADE} and therefore in the linkage between aerosols and the SWA atmospheric dynamics. It is only valid for the denoted spatiotemporal context of this modeling study and in the interval of F [0.1 1.5]. Further model realizations are necessary for more robust conclusions.

The NLLJ as a Large-Scale Atmospheric Feature over SWA

The findings of this section can support the interpretation of the observations at the supersites regarding the question whether the NLLJ develops locally or is a result of the EMFE front with its high postfrontal wind speeds around 250 m as the height where the NLLJ axis is expected (Fig. 5.9). From the modeling perspective there is clear evidence for a NLLJ that develops via the EMFE front for the night from 2 to 3 July entirely over SWA. For example the meridional vertical transect of wind speed capturing Savè shows no signal of a NLLJ before the passage of the EMFE front (not shown). To emphasize the strong gradient in wind speed and temperature, that can be found at the EMFE front, Figure 6.31 presents a meridional vertical transect for 5.75°W that is the same transect as used for the detailed analysis in Figure 6.24 (left). The color shading denotes the wind speed and the contours the potential temperature for the reference case (Fig. 6.31a) and the clean case (Fig. 6.31b). The EMFE front (vertical dashed line), that is detected by using the 302 K isentrope (thick black line), is ahead in the clean case compared to the reference case as described above. Figure 6.31 clearly shows that the postfrontal NLLJ is a large scale phenomenon, affecting the entire domain up to the coast. In contrast, the prefrontal wind speeds are low. The strong jump of the potential temperature at the front reveals cold air advection related to EMFE. The flow patterns are similar to that of a density current where fast moving cold air and surface friction lead to the formation of an elevated leading edge (overhanging nose), which can extend to higher altitudes than the tail (Simpson, 1969; Sun et al., 2002).

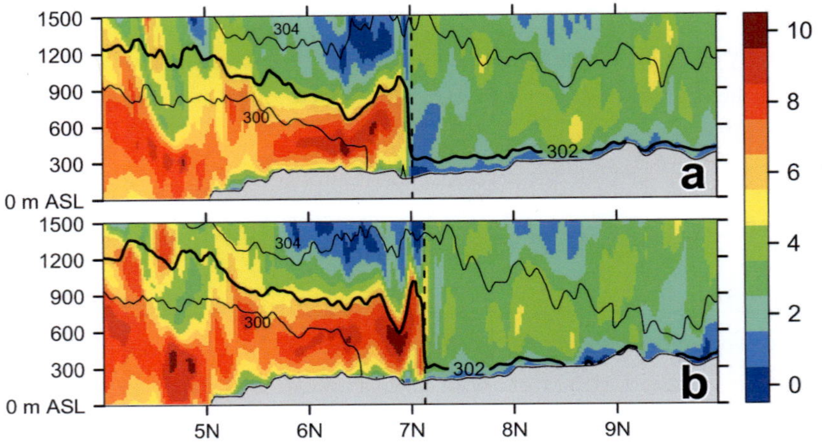

Figure 6.31: Meridional vertical transect (m ASL) of wind speed (shading, m s^{-1}) along 5.75°W (central Ivory Coast) for 2 July 21 UTC (compare with the left part of Fig. 6.24) for (a) the reference case and (b) the clean case. The black contour denotes the potential temperature (K) while the thick isentrope (302 K) is used for the identification of the EMFE front (vertical dashed line). The gray shading denotes the topography.

Vertical extensions of the nose of about 1 km are found for atmospheric density currents (Simpson, 1969), which agrees with the flow in Figure 6.31. Sun et al. (2002) emphasize that the wind surge behind the nose, that propagates close to the ground, leads to strong turbulent mixing.

6.3.3 Aerosol Impact on EMFE-Related Clouds and Precipitation

This section aims on the quantification of the cloud and precipitation characteristics related to the EMFE front. Furthermore, the changes in clouds and precipitation that relate to the different aerosol scenarios directly and (with the findings of Section 6.3.1 and 6.3.2) indirectly via the EMFE frontal shift will be analyzed.

Spatial Distribution of Convective Clouds and Precipitation

Figure 6.32 shows the precipitation and total cloud water for the clean case (Fig. 6.32a-b) and the reference case (Fig. 6.32c-d) over Ivory Coast. The precipitation (total cloud water) is accumulated (averaged) over the time period

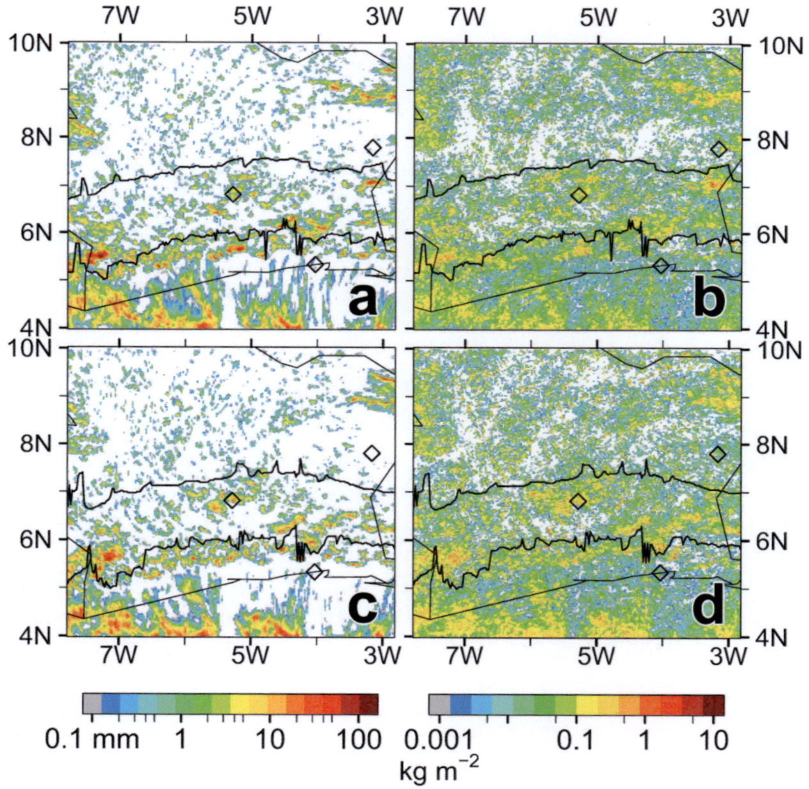

Figure 6.32: Accumulated precipitation (left, mm) and mean total cloud water (right, kg m^{-2}) for the time period 2 July 15-22 UTC over Ivory Coast for the clean case (a,b) and reference case (c,d). The black lines denote the location of the EMFE fronts at 15 UTC (south) and 22 UTC (north).

2 July 15-22 UTC. The precipitation postfrontal of the 15 UTC EMFE front (Fig. 6.32a,c) appears localized and along the front like in a rope of pearls, reaching high amounts of up to 100 mm. The predominant precipitation area is enveloped by the EMFE fronts at 15 UTC and 22 UTC. The prefrontal area of the 22 UTC EMFE front is only affected by sparse convective precipitation with weaker rainfall intensities. These findings correspond to the spatial distibution of total cloud water. The differences between the clean and reference case are less distinct. Qualitatively, it can be seen that the peak rainfall amounts are stronger and the total cloud water values are smaller in the clean case (Fig. 6.32a,b) compared to the reference case (Fig. 6.32c,d). The differences will subsequently be quantified.

Significance of the Precipitation Response on Aerosol

To assess whether the different aerosol burden in the clean and reference case leads to statistically significant differences in the EMFE front-related spatial precipitation pattern, the hatched blue area in Figure 6.33 was selected for further analysis. This area is limited to the South by the coastline and to the North by the area that is crossed by the EMFE front jointly in the clean and reference case between 15-22 UTC on 2 July (compare Fig. 6.32). The precipitation was accumulated over this time period capturing the hatched blue area for both realizations, leading to two datasets with the mean and variance presented in Table 6.2. Values below 0.1 mm in the accumulated precipitation fields are omitted.

Precipitation generally follows the gamma distribution and therefore the question about the statistical significance of the precipitation difference can be translated to the question whether the differences in the characteristics of the gamma distribution, representing the precipitation, are statistically significant. The gamma distribution is described via the shape parameter k and the scale parameter Θ. These parameters can be derived with the following relations for the first two moments of the gamma distribution:

$$E(x) = k\,\Theta \tag{6.2}$$

and

$$s^2(x) = k\,\Theta^2, \tag{6.3}$$

where $E(x)$ is the expected value and s^2 the variance. Expected value and variance are estimated via the mean $\bar{x}$ and variance s^2 of the datasets. The resulting parameter estimations for k and Θ are given in Table 6.2. The assumption to express the precipitation data via a gamma distribution and via the selected parameters is supported by the outcomes of quantile-quantile plots comparing the precipitation from the COSMO-ART perspective with the gamma distribution perspective (not shown). The uncertainty in the estimations of k and Θ is assessed via the bootstrapping technique using 10.000 samples. From these samples, the confidence intervals of the estimated parameters can be derived, in this case with a gaussian approximation and by using a significance level α=0.05. In Table 6.2, the parameters of the gamma distribution are given together with their confidence intervals.

Table 6.2: Statistical properties of the accumulated precipitation capturing 2 July 15-22 UTC over the area denoted in Figure 6.32a, for the clean and reference case. The parameters k and Θ denote the shape and scale parameter of the gamma distribution, respectively, describing the modeled precipitation patterns. The confidence intervals are related to a significance level of α=0.05.

Realization	**Mean $\bar{x}$ (mm)**	**Variance s^2**	**k**	**Θ**
$AIE_{0.1}ADE_{0.1}$	2.62	27.82	0.25±0.03	10.62±1.64
$AIE_{1.0}ADE_{1.0}$	2.51	16.47	0.38±0.03	6.55±0.54

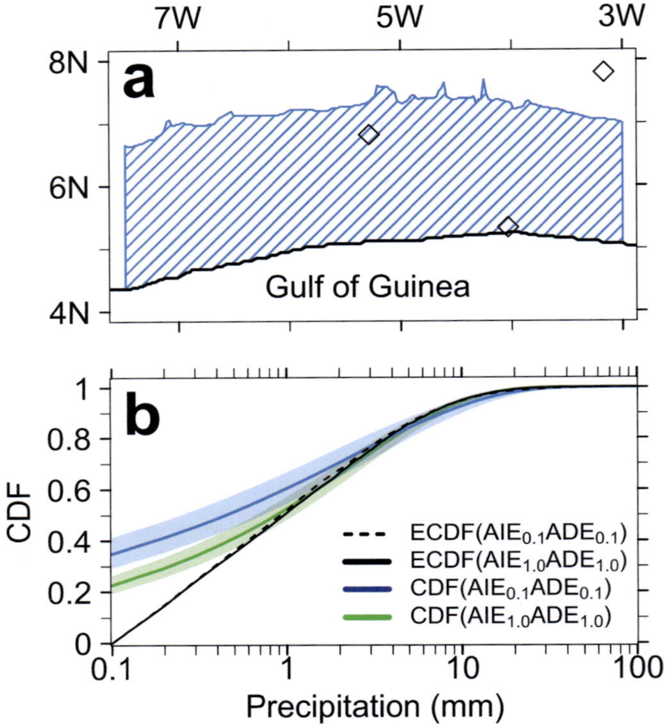

Figure 6.33: (a) Joint area that is crossed by the EMFE front in the clean and reference case on 2 July between 15 and 22 UTC (hatched blue), to the south extended to the coastline (black line). For this area the precipitation was accumulated for 2 July between 15 and 22 UTC for statistical analysis. (b) Cumulative distribution function (CDF) for the COSMO-ART perspective (Empirical CDF) clean (dashed black) and reference case (black solid) and for the gamma distribution perspective clean (blue) and reference (green). For the latter the uncertainty from the parameter estimate is added as shading. Accumulated precipitation values below 0.1 mm were omitted. Consider the logarithmic precipitation scale.

Although the parameters of the gamma distributions are significantly different between the clean and reference case, the cumulative distribution function (CDF) in Figure 6.33b reveals that the slope of the CDF is not significantly different between the realizations. The dashed and black solid lines denote the clean and reference case, respectively, based on the COSMO-ART data (Empirical CDF). Blue and green denote the clean and reference case, respectively, based on the estimated gamma distributions. For precipitations above 1 mm, the gamma distributions represent the data well. Below 1 mm the probability is overestimated when using the gamma distributions. This is partly related to the neglect of precipitation values below 0.1 mm. By using the parameter confidence intervals, envelopes are added to the CDFs to characterize the uncertainty. Within this uncertainty no significant difference is visible between the CDFs of the clean and reference case. The difference between the ECDFs are small but a slightly higher (lower) probability for precipitation values below 10 mm (above 10 mm) is visible for the clean case compared to the reference case.

From Table 6.2 it is obvious that the spatially averaged precipitation is virtually equal between the realizations. Furthermore, the reference case is related to a reduction in the precipitation area of about 13.1 % within the blue area of 6.33a. These two aspects, together with the differences in the ECDF, indicate that the precipitation is more localized in the reference case and over these local precipitation areas the rainfall intensity is higher since the mean precipitation is similar in both realizations. This precipitation localization with higher aerosol loading was also denoted by Saleeby et al. (2014) in a study of shallow tropical convection. However, as shown above, these effects are not statistically robust.

In Figure 6.34 the differences in the cloud properties on 2 July are quantified via number size distributions of the cloud droplets. The droplet size is characterized by the effective radius. Figure 6.34a-c shows the results for the reference case at 15, 18 and 21 UTC and Figure 6.34d-f for the clean case at 15, 18 and 21 UTC over the area denoted in Figure 6.33a. A clear signal of reduction in the CDNC with a reduction in the aerosol number is visible that is linked with an increase in the cloud droplet effective radius. The gray horizontal and vertical lines show the median value for the number concentration and the effective radius, respectively. The reduction of the aerosol number in the aerosol activation by one order of magnitude from 1.0 to

0.1 (reference to clean) leads to a median reduction of the CDNC from 100-200 to around 50 cm^{-3} and to an increase in the effective radius from 8 to 11 μm. The results underline that the aerosol factor change from 1.0 to 0.1 significantly alters the cloud droplet spectrum.

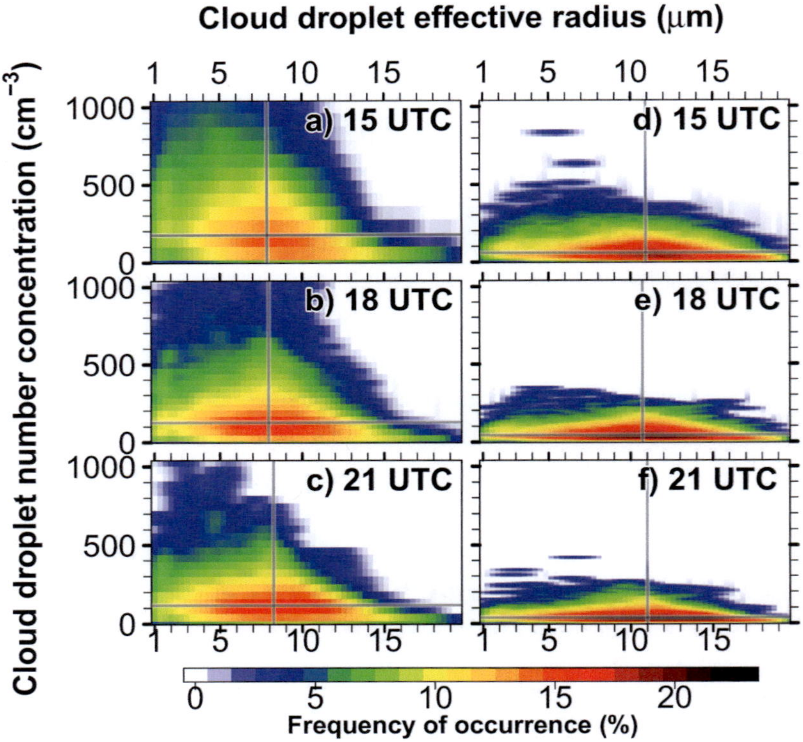

Figure 6.34: CDNC as diurnal cycle for 2 July 15-21 UTC over the area denoted in Figure 6.33a for (a) the reference case and (b) the clean case, with the cloud droplet effective radius (μm) and CDNC (cm^{-3}). The gray lines denote the median of the cloud droplet effective radius and CDNC. The color shading shows the frequency of occurrence (%).

EMFE Impact on the Spatial Precipitation Distribution

To assess whether clouds and precipitation are also affected by the EMFE meridional shift, Figure 6.35 shows the precipitation displacement relative to the EMFE front location. The precipitation is averaged zonally over Ivory Coast for the 60 km EMFE pre- and postfrontal area for 2 July 15-22 UTC. The EMFE front and the precipitation of the reference case is centered to zero (black) and the clean case is shown with its relative meridional shift (blue). The results again underline that the postfrontal area is linked with higher precipitation amounts than the prefrontal area (compare Fig. 6.32a,c). The prefrontal precipitation results from the thermal convection that diminishes after sunset and the postfrontal precipitation to the convergence and vertical lifting at the EMFE front. At 15 UTC and less pronounced at 16 UTC (Fig. 6.35a-b) a minimum of precipitation is visible at the EMFE front location and directly postfrontal. This feature indicates that at this time the EMFE front is still in its stationary phase. During that time the impediment of the monsoon flow through the turbulence over land lead to postfrontal downward vertical motion that can clearly be seen in meridional transects of wind speed between 12 UTC and 15 UTC (not shown). This postfrontal downward motion is in agreement with the frontal study over Mauritania of Grams et al. (2010). The downward motion in the postfrontal meridional area of 10-15 km leads to a cloud dissolution and suppression of precipitation. After 15 UTC the turbulence barrier diminishes and the EMFE front can propagate inland. Therefore the downward motion transforms to a horizontal motion and precipitation can also develop in direct vicinity of the front. The reference case generally tends to produce less precipitation than the clean case (Albrecht effect) and the frontal differences follow the analysis of Section 6.3.1. At 15 UTC and 16 UTC, characterized by the stationarity of the EMFE fronts, the precipitation in the clean and reference case have the same meridional position. After 16 UTC, with the faster inland propagation of the EMFE front in the clean case, a clear meridional precipitation shift is visible.

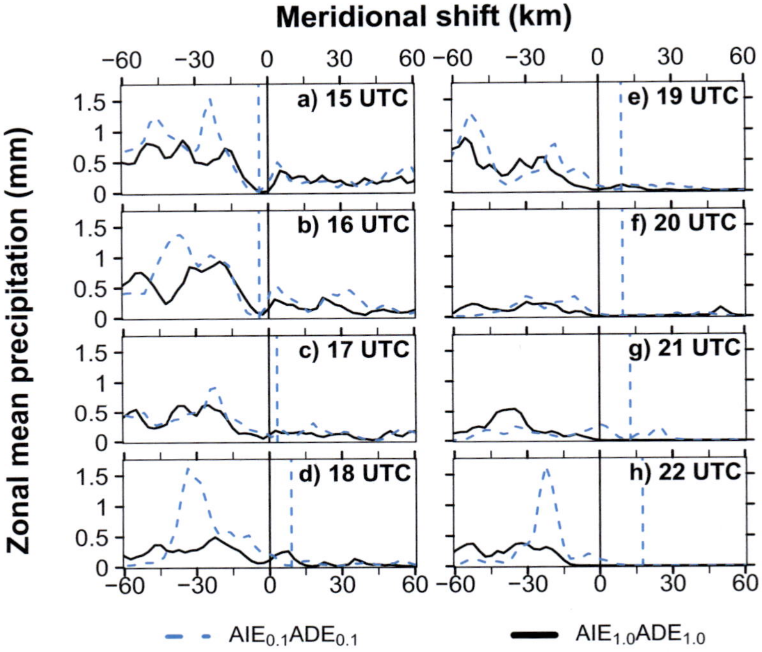

Figure 6.35: Zonal mean precipitation (mm) as diurnal cycle for 2 July 15-21 UTC over Ivory Coast. The precipitation was zonally averaged along the longitudinal course of the EMFE front in the reference case (black solid) and the clean case (blue dashed), respectively. To allow for a comparison of the zonal precipitation shift (km) due to the shift of the EMFE fronts in the different realizations, the EMFE front for the reference case (vertical black solid line) is centered to zero. By considering the relative difference between the location of the EMFE front in the reference and clean case, the precipitation pattern of the clean case is added together with the location of the front (vertical blue dashed line). For the comparison of the spatial precipitation shift, the area 60 km ahead and behind the reference case EMFE front is shown. It has to be highlighted that the panels show the spatial shift relatively according to the EMFE front locations. The panels therefore does not imply a stationarity of the reference case EMFE front. Both fronts are propagating inland with time. Zero values of precipitation were omitted.

With the advance of the EMFE front also the precipitation is advanced towards the reference case. At 21 UTC (Fig. 6.35g), a postfrontal precipitation shift of up to 15 km can be observed for the clean case.

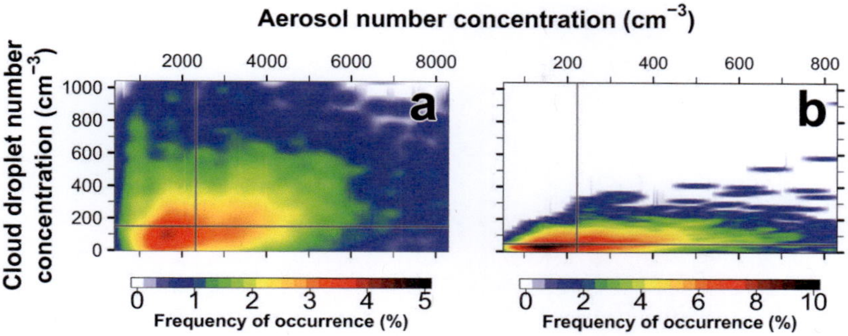

Figure 6.36: Scatter plot of total aerosol number concentration as applied for the aerosol activation (cm^{-3}) and CDNC (cm^{-3}) capturing the 20 km EMFE postfrontal area over Ivory Coast (7.5-3.0°W) for (a) the reference case and (b) the clean case on 2 July 15 UTC by omitting values below 1 cm^{-3}. The gray lines denote the median of the aerosol number concentration and CDNC. The color shading shows the frequency of occurrence (%). Consider the different abscissas.

Figure 6.35 underlines that the change in the EMFE front due to aerosol is furthermore a pathway to create more precipitation further inland.

The relationship between the total number of aerosols and the CDNC in the 20 km EMFE postfrontal area on 2 July 15 UTC is presented in Figure 6.36 for the reference and clean case. The median CDNC of 100-200 cm^{-3} for the reference case (Fig. 6.36a) and around 50 cm^{-3} for the clean case agree with the findings of Fig. 6.34. In the reference (clean) case these droplet number concentrations are related to aerosol number concentrations of around 2000 (200) cm^{-3}. As a comparison, in the TMMS without considering prognostic aerosol, an aerosol number concentration of 1700 cm^{-3} is assumed for continental conditions. Martin et al. (2010) measured aerosol number concentrations of up to 10.000 cm^{-3} related to biomass burning in the Amazon. The number concentrations in Figure 6.36b denote a rather pristine atmosphere. In addition, Figure 6.37 shows the temporal evolution of the total cloud water in the same area considered in Figure 6.36. With the decrease in the aerosol number concentration in the clean case also the total cloud water generally decreases due to precipitation loss, following the Albrecht effect.

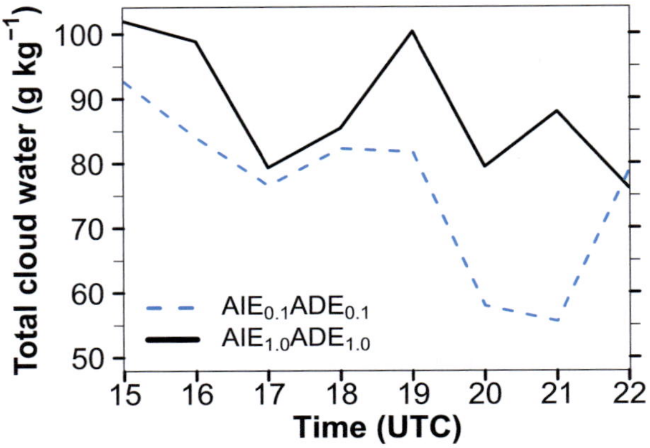

Figure 6.37: Temporal evolution of total cloud water (g kg^{-1}) averaged meridionally over the 20 km EMFE postfrontal area of the clean case (blue dashed) and reference case (black solid) and zonally over Ivory Coast (7.5-3.0°W) for 2 July 15-22 UTC by omitting zero values.

This section has shown that the entire SWA domain is affected by the EMFE front, a front that is located stationary over land near the coast between midday and the early afternoon and that propagates inland after 15 UTC, largely contributing to the observed precipitation. It turnes out that the EMFE front is highly sensitive to aerosol-induced changes via differences in the near-surface heating and a susceptible balance between the monsoon flow and inland turbulence. A stronger pressure gradient force in the clean case intensifies the acceleration of the EMFE front, leading to inland shifts of the EMFE front accompanied by inland precipitation shifts compared to the cases with higher aerosol amounts. The EMFE postfrontal precipitation amount shows no significant difference betwen the clean and the reference case The EMFE frontal patterns and their aerosol-induced shifting are coherent over Ivory Coast. The aerosol factor (F_{ADE}, F_{AIE}) decrease from 1.0 to 0.1 substantially alters the cloud droplet spectrum with reduced CDNC and increased cloud droplet effective radii.

The subsequent section will change the focus to the characteristics of the NLLS. However, this denotes no clear cesura since the EMFE front and its response to aerosol are also affecting the low-level clouds.

6.4 Aerosol Affecting the Nocturnal Low-Level Stratus (NLLS)

As described in Section 2.2, SWA is frequently covered by an extensive deck of low-level stratus that evolves nocturnally (NLLS). This section quantifies the spatiotemporal evolution of the NLLS in the case study period, provides a detailed analysis of the cloud micro- and macrophysical properties for the Lomé-Savè transect and will assess potential feedbacks with respect to changes in the aerosol amount.

6.4.1 NLLS Spatiotemporal Overview

Whereas in Section 6.3 the focus was located on the time period 15-21 UTC for 2 July (and briefly also for 3 July), for the NLLS analysis the night and the morning (2 July 21 UTC to 3 July 10 UTC) are of relevance. When considering the stratus-to-cumulus transition, the time period is extended to 16 UTC.

Spatiotemporal Distribution of Low-Level Clouds

Figure 6.38 provides an overview of the temporal evolution of the low-level cloud spatial distribution of the denoted time frame within the reference case. Only selected times from the 14 h time period are shown to describe the characteristic features. The low-level clouds are shown as blue shading, defined by nonzero cloud water in the lowest 1.3 km AGL. A topographic height above 250 m ASL is shown in brown.

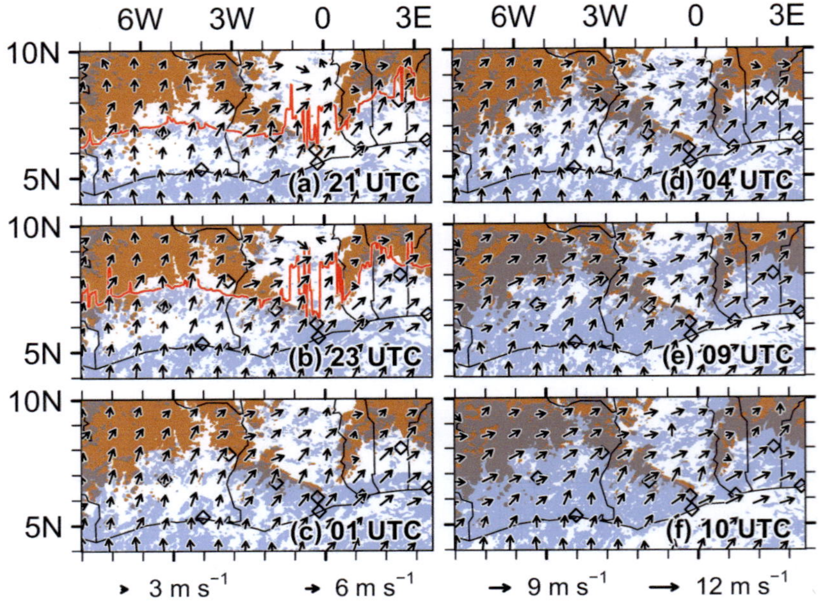

Figure 6.38: Overview of the low-level cloud temporal evolution between 2 July 21 UTC and 3 July 10 UTC (a-f) for the reference case. Blue shading denotes low-level clouds via the existence of cloud water in the lowest 1.3 km AGL. Brown shading indicates the topography above 250 m ASL. The arrows show the wind speed (m s^{-1}, scale is given below) and direction at 250 m AGL. For 21 UTC and 23 UTC the EMFE front for the reference case is shown in red.

Additionally, the wind vectors describe wind speed and direction at 250 m AGL. 21 UTC (Fig. 6.38a) denotes the end of the time period of the EMFE analysis in Section 6.3. The red line shows the location of the EMFE front. A 1° zonal cloud area in the EMFE postfrontal area is visible for the entire domain except of the Lake Volta area. Between this cloud band and the coastline, a remarkable low coverage with low-level clouds is visible. The offshore area is densely covered by low-level clouds. Ivory Coast is affected by southerly winds and to the east, the wind direction gradually turns to southwest. The Mampong Range and the Atakora Mountains contribute to this change in wind direction. Two hours later at 23 UTC (Fig. 6.38b) an increase in the spatial coverage with low-level clouds can be seen, on the one hand by increased cloudiness in the EMFE postfrontal area and on the other hand by the inland penetration of clouds from the Gulf of Guinea. The latter is obvious for the entire coastline from Ivory Coast to Benin. Over Ivory Coast this development leads to two zonal cloud bands (postfrontal and coastline) that are separated by a rather cloud-free area in between. Over Ghana these two cloud bands are connected via extensive low-level clouds at the Mampong Range, very likely related to lifting and blocking at the topographic barriers. Over Togo and Benin also this double band structure is visible but less pronounced. Again two hours later at 1 UTC (3 July, Fig. 6.38c) an extensive low-level cloud area has established upstream of the entire length of the Mampong Range in Ghana. Also the advance in the inland propagation of the coastal low-level clouds is remarkable, especially in the eastern part of the domain. The cloud band in this area is originating from clouds around 1°E that are advected in northeasterly direction over Accra and Ghana to the coastal area of Ghana, Benin and western Nigeria. However, the clouds are not only advected but they also intensify and expand with the inland propagation.

At 4 UTC (Fig. 6.38d) the low-level cloud deck further expands. East of the Atakora Mountains the cloud coverage is denser than over Ivory Coast. The surroundings of Savè are still cloud free but from the South the coast-related low-level clouds and from the North the clouds from the Oshogbo Hills approach, uniting over Savè around 8 UTC (not shown). From 4 UTC (Fig. 6.38d) to 10 UTC (Fig. 6.38f) the wind direction over Ivory Coast changes from south to southwest and east of the Atakora Mountains to west-southwest. At 9 UTC (Fig. 6.38e), that is identified by many studies as the time with the maximum NLLS expansion (e.g. Schuster et al., 2013), also the COSMO-ART realization shows a high cloud coverage. At Savè supersite a significant increase in the NLLS density was observed after sunset (5:43 UTC) since the onset of surface fluxes leads to a moisture boost from the lower layers (B. Adler, personal communication).

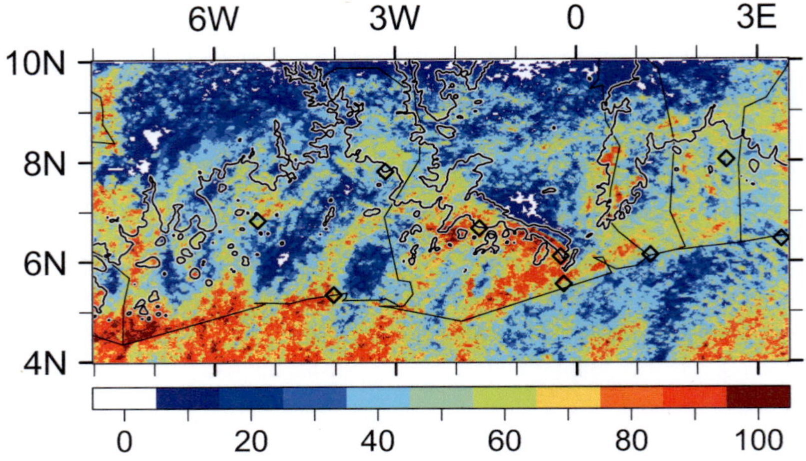

Figure 6.39: Frequency of occurrence (%) of low-level clouds in the time period 2 July 21 UTC to 3 July 10 UTC for the reference case. The black contour denotes the topography above 250 m ASL.

This agrees well with the modeled cloud cover in Figure 6.38. The maximum expansion in the reference case is reached at 10 UTC (Fig. 6.38f), capturing nearly the entire land area from the coastline to 10°N. Again the Lake Volta Basin north of the topographic barriers is an exception with less low-level clouds. Nevertheless, this area is not cloud free but affected by clouds at higher altitudes (not shown). The strong impediment of the cloud propagation at the Mampong Range leads to much denser clouds over southwestern Ghana than in the other regions. The cloud area, visible between Lomé and Savè, will be subject to a detailed analysis since for this area aircraft observations are available and the interactions between the NLLS and the air pollution from Lomé can be investigated. Figure 6.39 summarizes the findings obtained above by presenting the frequency of occurrence of low-level clouds in the 14 h time period from 2 July 21 UTC to 3 July 10 UTC. As expected, the occurrence frequency for low-level clouds is highest over the ocean and at topographic barriers, in particular the Mampong Range. In the lee of the Mampong Range a distinct minimum of low-cloud occurrence is visible. It is supposed that this is partly related to the blocking of clouds via the mountain range, the lifting of moisture and clouds to greater heights and to a Föhn-related cloud dissolution (Schuster et al., 2013). In the eastern area, enveloped by the Atakora Mountains and Oshogbo Hills, the occurrence is homogeneously at or above 40 %, whereas over Ivory Coast and southwestern Ghana alternating band-like areas with less and high frequencies occur. To exclude non-physical artefacts from the nesting process, the coarser domain D2 was inspected visually. It turnes out that these cloud bands are widespread over the eastern equatorial Atlantic Ocean, advected with the southerly monsoon winds. This is also confirmed by the clouds observed by MODIS Terra around 11:11 UTC (Fig. 6.40a). When taking these band-like structures into consideration for the low-level cloud formation over Ivory Coast between 2 July 21 UTC and 3 July 10 UTC, it is clearly visible that these cloud bands propagate northward onshore, conserving the alternating pattern of cloudy and cloud-free areas. The gaps between the cloud bands start to close after sunrise. A closed cloud deck is visible at 10 UTC. The finding from the model results, that the inland cloud band over the eastern part of the domain results from a northeastward advection of clouds from the offshore region of Ivory Coast, is in agreement with Figure 6.40a in which the Gulf of Benin (bottom right corner of the red rectangle) is significantly less affected by clouds that are advected from the South than the offshore area of Ivory Coast (bottom left corner of the red rectangle). At around 11 UTC the remote sensing data reveals already a stratus-to-cumulus transition that is more pronounced over Ivory Coast and southwestern Ghana than over the eastern part of SWA (Fig. 6.40b). The eastern stratus and stratocumulus also appear much thicker and denser than the clouds over Ivory Coast. As indicated by Figure 6.38 and 6.39, the most dense clouds can be found upstream of the Mampong Range.

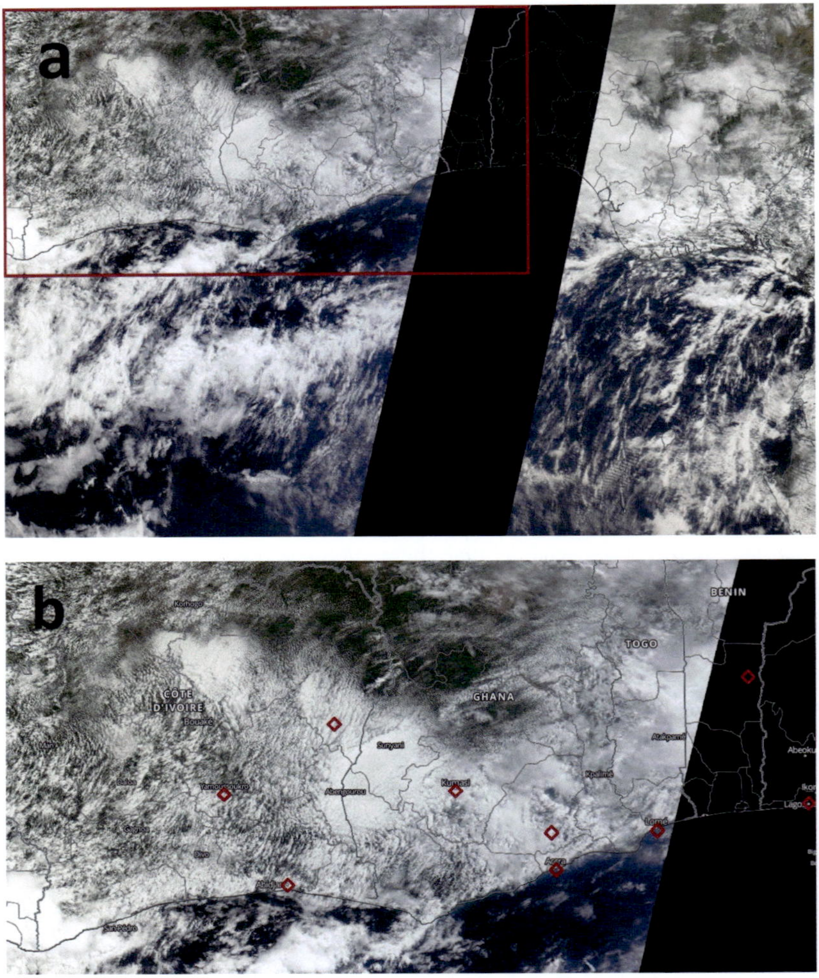

Figure 6.40: Cloud cover observed by MODIS Terra (NASA, 2017) on 3 July 11:11 UTC for (a) SWA and the eastern equatorial Atlantic Ocean and (b) focused on SWA by capturing the area denoted by the red rectangle in (a). The red diamonds indicate the location of SWA cities as shown in the model result figures.

Spatiotemporal Microphysical Characteristics of the NLLS

Figure 6.41 shows the vertically integrated cloud water for the total column (Fig. 6.41a) and the lowest 1.3 km AGL (Fig. 6.41b) on 3 July 9 UTC. The modeling results support the MODIS observations by showing higher cloud water values over extensive areas in the eastern part than over central Ivory Coast.

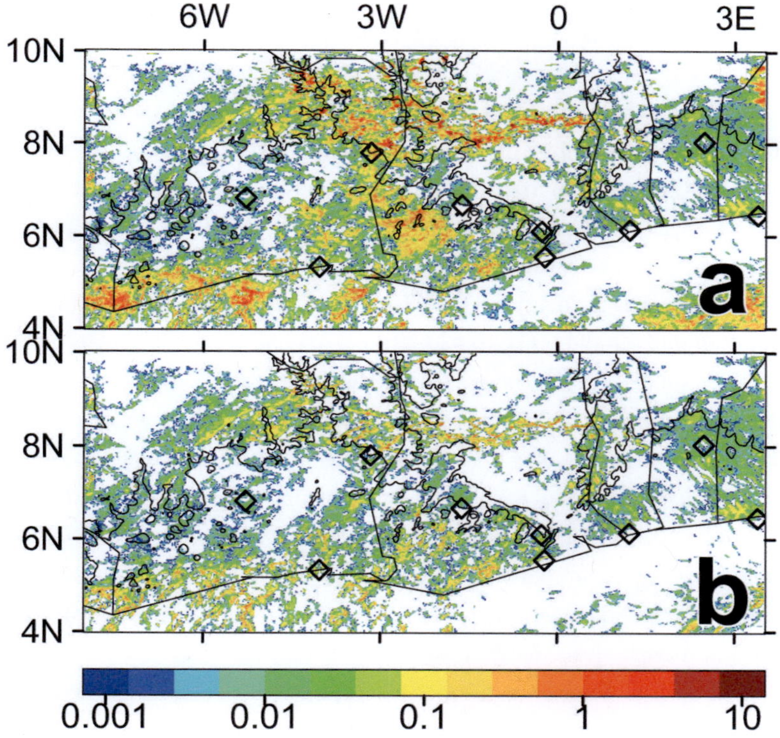

Figure 6.41: Vertically integrated cloud water (kg m^{-2}) on 3 July 9 UTC for (a) total vertical column and (b) lowest 1.3 km AGL. The black contour denotes the topography above 250 m ASL.

Also the extensive clouds with a high cloud water content upstream of the Mampong Range are in agreement with the cloud distribution observed by MODIS. From the comparison of Figure 6.41a and 6.41b it can be seen that a large cloud water contribution for the Mampong Range clouds is related to heights above 1.3 km AGL. The DACCIWA supersites (in particular Savè) provide a comprehensive dataset that characterize the evolution of the NLLS. Nevertheless, the scope of this section is to assess the sensitivity of the SWA meteorological characteristics on changes in aerosol rather than the evaluation of COSMO-ART with available meteorological observations provided at Savè supersite, as it was done for the D1 forecasts. Although Savè is not in the key focus of our study in terms of aerosols, a detailed evaluation of D3 against meteorological measurements in a follow-up study could contribute added value to the understanding of the COSMO-ART performance.

Further insight in the microphysical characteristics of the NLLS is achieved by Figure 6.42 and 6.43 for 3 July 4 UTC and 9 UTC, respectively. For the height of the cloud water maximum in the lowest 1.3 km AGL (Fig. 6.42a and 6.43a) the cloud water (Fig. 6.42b and 6.43b), CDNC (Fig. 6.42c and 6.43c), cloud droplet effective radius (Fig. 6.42d and 6.43d) and the total aerosol number concentration (Fig. 6.42e and 6.43e) are shown.

For 4 UTC it is clearly visible that the height of the cloud water maximum is lowest for the zonal NLLS bands east of the Atakora Mountains and over Ivory Coast (200-400 m AGL). Upstream of the Mampong Range, over the Lake Volta Basin and over the ocean the cloud water maximum is located significantly higher around 900-1300 m AGL (Fig. 6.42a). The NLLS is linked to maximum cloud water values around 0.05-0.3 g kg^{-1} (Fig. 6.42b), CDNC of 100-200 cm^{-3} (Fig. 6.42c) and effective radii of 7-10 μm (Fig. 6.42d). The total aerosol number concentration shows, as expected, a strong contrast between the rather clean marine area (500-1000 cm^{-3}) and the inland area that is affected by high aerosol numbers around 5000-10.000 cm^{-3}. Clearly visible is the pollution plume downstream of Lomé located in the connecting line to Savè. Also the pollution plumes of Abidjan and Accra are indicated.

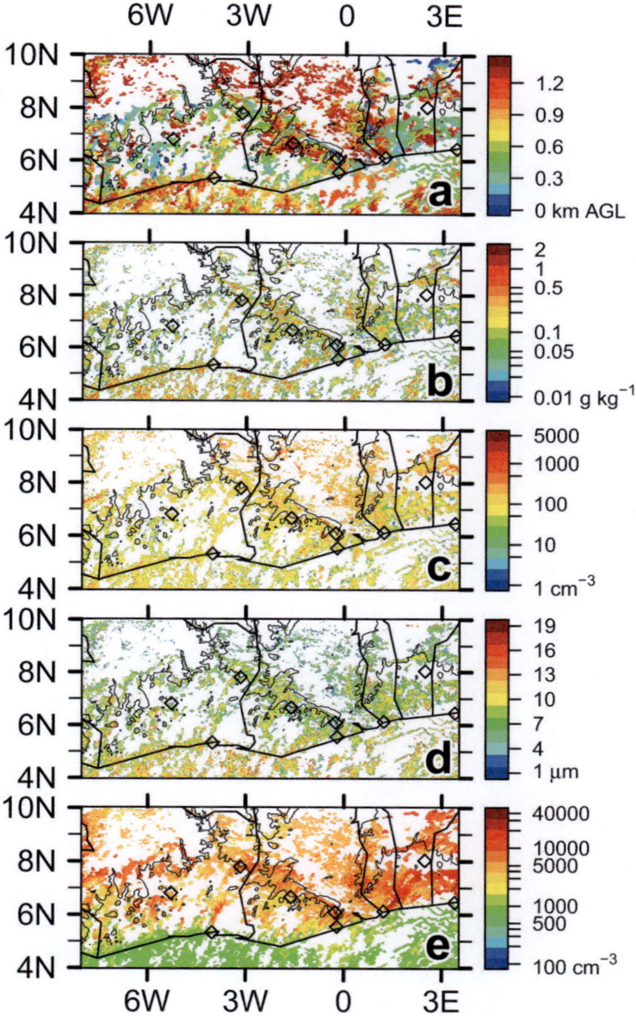

Figure 6.42: Aerosol and cloud characteristics on 3 July 4 UTC for the height AGL of the cloud water maximum in the lowest 1.3 km AGL as presented in (a). The subsequent panels show (b) the cloud water (g kg^{-1}), (c) the CDNC (cm^{-3}) (d) the cloud droplet effective radius (μm) and (e) the total aerosol number concentration (cm^{-3}). The black contour denotes the topography above 250 m ASL.

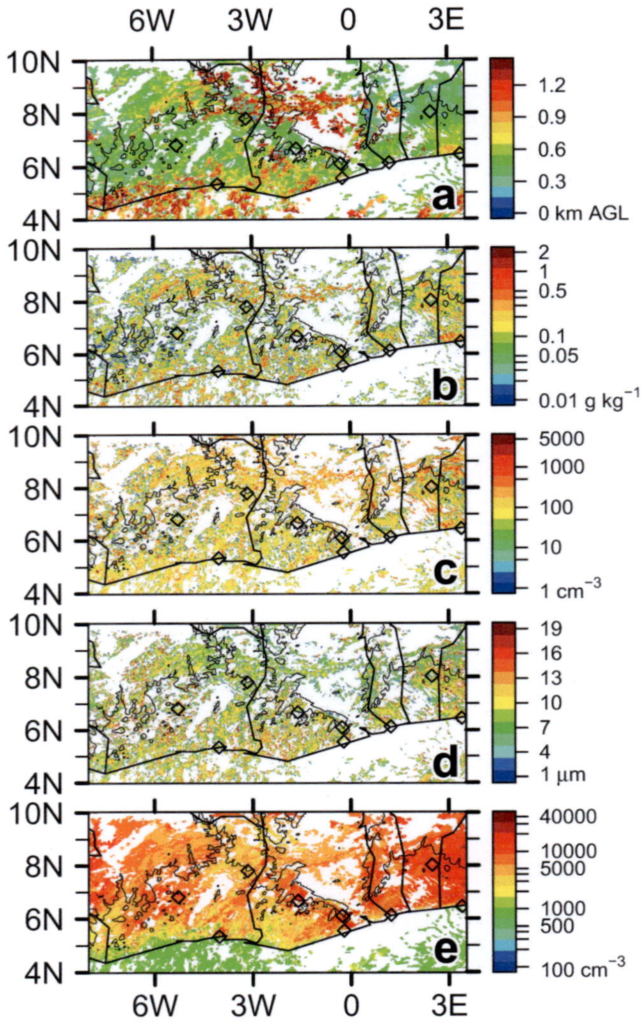

Figure 6.43: Same as Figure 6.42 but for 3 July 9 UTC.

At 9 UTC the height AGL of the cloud water maximum is very homogeneous over wide areas of SWA around 400-500 m AGL (Fig. 6.43a) and indicates already a vertical lifting compared to 4 UTC. For the upstream area of the Mampong Range the height is around 500-700 m AGL. At 4 UTC an extensive cloud area upstream of the Oshogbo Hills has developed that captures also Savè. Compared to the NLLS south of Savè at 4 UTC, this cloud area shows higher cloud water values (0.1-0.3 g kg^{-1}, Fig. 6.43b), a slightly higher CDNC (100-300 cm^{-3}, Fig. 6.43c) and higher effective radii (around 10 μm, Fig. 6.43d). From Figure 6.42e and 6.43e it can be deduced that the downstream area of Lomé (at least south of 7°N) is likely not affected by the emissions of Accra, since the two pollution plumes are rather parallel. This becomes important when interpreting the NLLS characteristics of the Lomé-Savè area in the subsequent sections.

Displacement of the Inland Propagating NLLS via the Aerosol-EMFE Feedback

To qualitatively assess whether a displacement of the inland propagating NLLS due to the aerosol-related changes in the EMFE takes place, Figure 6.44 shows the spatial distribution of clouds below 1.3 km AGL for 2 July 22 UTC (Fig. 6.44a) and 3 July 2 UTC (Fig. 6.44b).

Red (blue) areas denote clouds only in the reference case (clean case) whereas gray areas represent clouds that exist in both realizations. At 22 UTC a clear advance of the inland propagation of clouds from the Gulf of Guinea can be observed for the clean case over Ivory Coast. This advance qualitatively corresponds to the magnitude of the EMFE frontal shift, denoted by the two black lines over Ivory Coast with the clean case EMFE front ahead.

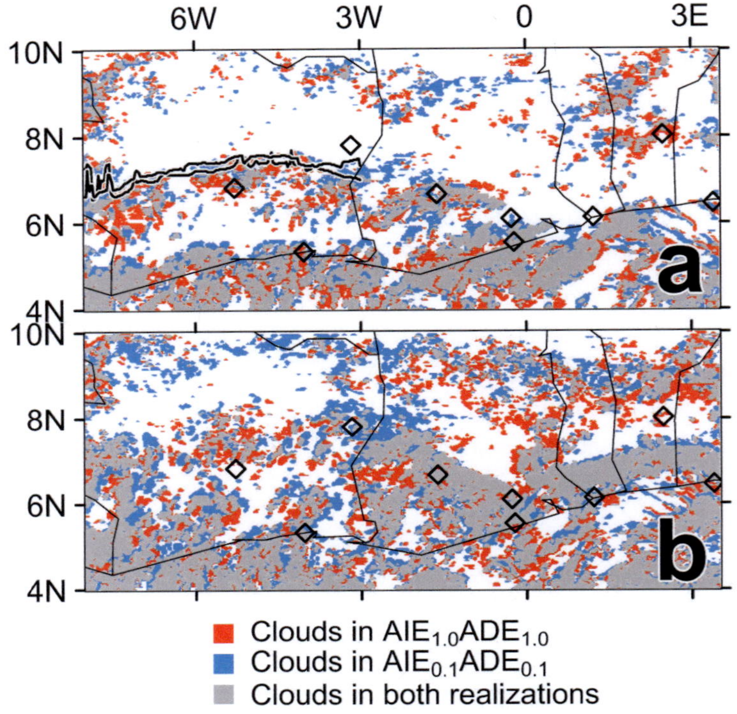

Figure 6.44: Spatial cloud coverage considering the lowest 1.3 km AGL for (a) 2 July 22 UTC and (b) 3 July 2 UTC. The red and blue colors denote areas that are only affected by clouds in the reference (clean) case. For the gray areas both realizations simultaneously show cloud cover. The black lines in (a) show the EMFE front over Ivory Coast for the reference and clean case with the latter ahead.

Additionally, at 2 UTC the advance of the zonal cloud band in the eastern part of the domain is visible for the clean case. The differences will not be further investigated but from a qualitative perspective the EMFE frontal shift also leads to a comparable shift in the inland propagation of coastal clouds.

6.4.2 Modeled Cloud Properties in Comparison with Aircraft Observations

The Twin Otter Research Flight on 3 July

For 3 July a research flight of the Twin Otter took place as part of the DACCIWA aircraft campaign (flight number TO-02). This flight was dedicated to the research topics: stratus clouds and Lomé city emissions. The flight track is shown in Figure 6.45a with the flight direction highlighted by arrows.
The DACCIWA research aircrafts had their flight base at Lomé airport. On 3 July the Twin Otter started from Lomé at 11:02 UTC, heading to Savè in northeastward direction. The aircraft position between 10:45-11:30 UTC, 11:30-12:30 UTC, 12:30-13:30 UTC and 13:30-14:06 UTC is shown in blue, gray, red and black, respectively. Figure 6.45b additionally shows the flight altitude with the same color coding as for Figure 6.45a. Additional information regarding TO-02 are available from the flight summary document provided by the aircraft scientist Keith Bower (University of Manchester, School of Earth and Environmental Sciences). This section refers to the flight summary document to support the interpretation of the findings.
As indicated by Figure 6.40b, Lomé (red diamond at the eastern edge of the available MODIS data) and in general the inland area are densely covered by clouds. The MODIS cloud image, valid for 11:11 UTC, fits with the time of the Twin Otter takeoff and location in the blue flight section of Figure 6.45a,b. After the takeoff at 11:02 UTC, the aircraft ascended to the cloud base around 300-400 m (blue section in Fig. 6.45b). Due to this low cloud base, the minimum safe altitude (600 m AGL in the Lomé area) was located in the cloud layer. An ascent through the cloud layer would therefore have meant that the aircraft cannot descend again. This is an important information for the subsequent interpretation of the cloud properties because the blue section does not include a cloud penetration. The first upward profile, denoting the transition from the blue to the gray section in Figure 6.45, reveals a

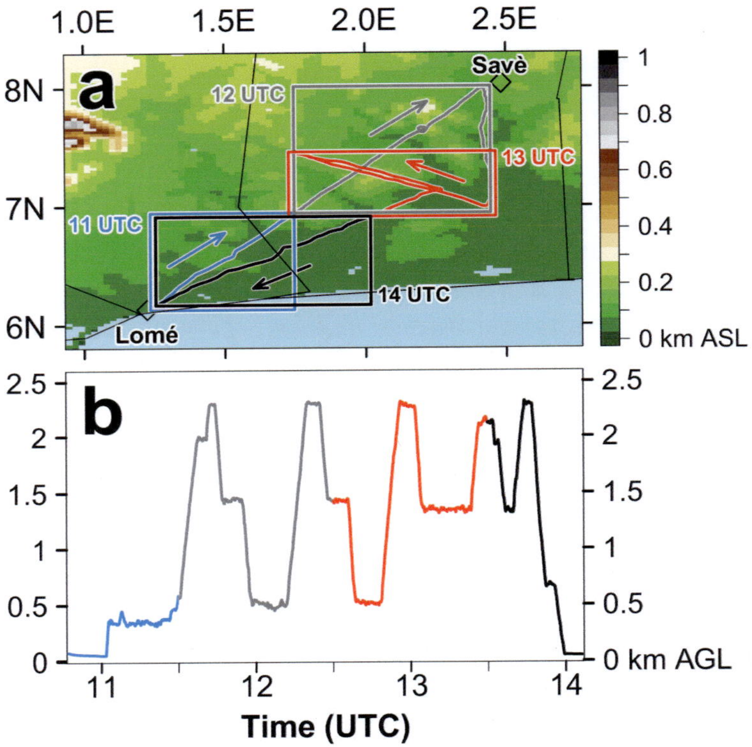

Figure 6.45: Scientific flight track of the Twin Otter aircraft on 3 July 2016 between 10:47 UTC and 14:06 UTC (flight number TO-02) in (a) horizontal and (b) vertical direction (km AGL). For (a) the topography (km ASL) is added. The flight track in (a) and (b) is separated in hourly time steps for the subsequent collocation with hourly model data from COSMO-ART, highlighted by the blue (10:47-11:30 UTC), gray (11:30-12:30 UTC), red (12:30-13:30 UTC) and black color (13:30-14:06 UTC). The rectangles, spanned by the horizontal extension of the hourly flight sections, are used for the selection of model data. Furthermore, the arrows in (a) indicate the flight direction with the takeoff at Lomé, the flight to Savè and the return to Lomé airport. Shortly after 12 UTC (with a flight altitude of 0.5 km AGL) the Twin Otter reached Savè. Consider the meridional compression of the map in (a).

cloud base of 730 m AGL and multiple embedded cloud tops (1200 m AGL and 1300 m AGL). It therefore can be concluded that at this location the thickness of the NLLS is approximately 500 m. The Savè supersite detected a cloud base of about 500 m AGL around noon as the time of the Twin Otter passage (compare Fig. 6.4). The second upward profile through the clouds was realized after passing Savè and turning southwards between 12:00 UTC and 12:30 UTC. For this area a cloud base, cloud top and cloud thickness of 380 m AGL, 730 m AGL and 350 m was observed, respectively, with additional clouds around 900 m AGL. Therefore this area exhibits a shallower and lower NLLS than observed in the southwestern area. With the location in the red section denoted in Fig. 6.45 around 13 UTC, the conditions are characterized as more convective with scattered convective clouds. So the scattered cumulus conditions, already observed at 11 UTC over central Ivory Coast by MODIS (Fig. 6.40b), have expanded to the eastern part of SWA by 13 UTC.

Boxplots of Modeled and Measured Cloud Properties

In the following, the CDNC, cloud droplet effective radius and cloud liquid water observed by the Twin Otter is used for a comparison with the COSMO-ART realizations. To avoid discrepancies from a mismatch in the observed and modeled cloud location, the comparison is not realized along the flight track but by using the horizontal areas that are spanned by the hourly flight track sections for 11-14 UTC (blue, gray, red and black rectangle in Fig. 6.45a, collocated to the hourly model output of COSMO-ART). In this case every rectangle is diagonally crossed by the aircraft observations. The observed and modeled quantities are compared via boxplots. Figure 6.46 shows the comparison with respect to the CDNC for 11, 12, 13 and 14 UTC (Fig. 6.46a-d) and considering the observations (gray boxplot) and the reference, clean and polluted case (green, blue and red boxplot, respectively). To exclude observations outside of clouds, the analysis is limited to CDNC above 10 cm^{-3}. For the interpretation it has to be considered that the different times also denote different areas.

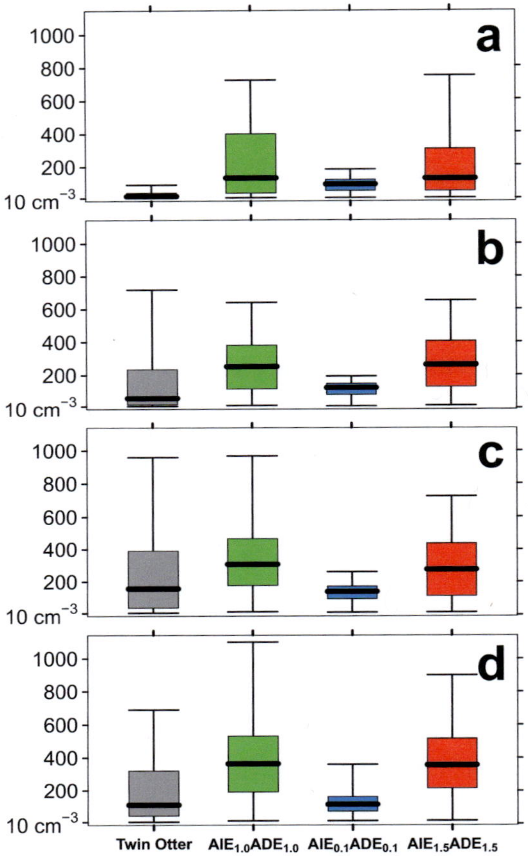

Figure 6.46: Boxplots of the CDNC (cm^{-3}) according to the flight track denoted in Figure 6.45a for (a) 11 UTC (blue flight track section and rectangle in Figure 6.45a), (b) 12 UTC (gray flight track section and rectangle in Figure 6.45a), (c) 13 UTC (red flight track section and rectangle in Figure 6.45a) and (d) 14 UTC (black flight track section and rectangle in Figure 6.45a). The gray boxplots show the Twin Otter observations (according to the flight track section, 1 s temporal resolution of the CDP device) and the green, blue and red boxplots show the model results for the reference, clean and polluted case within the corresponding rectangles, respectively. The whiskers capture the data from the 0.025 to the 0.975 quantile (95% of the data). Data outside this range are not shown. CDNC below 10 cm^{-3} were omited, leading to an observational data basis of 34, 443, 403 and 115 observations at 11, 12, 13 and 14 UTC, respectively. Regarding the model data only the lowest 2.3 km are considered in agreement with the Twin Otter maximum flight altitude.

At 11 UTC (Fig. 6.46a), the observed CDNC is negligible due to the above-mentioned fact that the Twin Otter stays below the cloud layer. At 12 UTC (Fig. 6.46b, corresponding to the gray section in Figure 6.45a), the observed median CDNC of 60 cm^{-3} is also small. The highest observed median CDNC is identified at 13 UTC (Fig. 6.46c, corresponding to the red section in Figure 6.45a). At 14 UTC (Fig. 6.46c, corresponding to the black section in Figure 6.45a) the median CDNC is smaller again. For the COSMO-ART realizations the CDNC are generally higher in the cases with more aerosol (green and red denoting reference and polluted case) than in the clean case (blue), as expected. The median CDNC of the reference and polluted case are comparable, indicating that for the reference case already a saturation of CCN is reached so that no more CCN are activated with the aerosol surplus in the polluted case. Interestingly, the clean case median CDNC generally fits better to the observed median CDNC than the reference case, although the variance in the clean case CDNC is much smaller than in the observations. In contrast to the observations, the model results clearly shows an increase in CDNC with time. Miles et al. (2000) collated a data base of observed cloud properties of low-level stratiform clouds. For example for the Madeira Islands they identified CDNC around 50 cm^{-3} for nocturnal stratus and around 300 cm^{-3} for cumulus and stratocumulus. Therefore it is expected that an ongoing stratus-to-cumulus transition leads to the increase in the modeled CDNC with time.

Additionally, Figure 6.47 shows the boxplot for the cloud droplet effective radius. In general the effective radii are between 5 and 7.5 μm. Highest values are reached in the clean case due to the limited availability of CCN. Especially at 13 and 14 UTC (Fig. 6.47c-d) the reference case is in good agreement with the observations.

Finally, Figure 6.48 shows the analysis regarding the observed and modeled cloud water. The results generally indicate a skewness in the distribution with many small values and less higher values. The median cloud water is continously below 0.2 g m^{-3} with a tendency towards higher modeled cloud water concentrations.

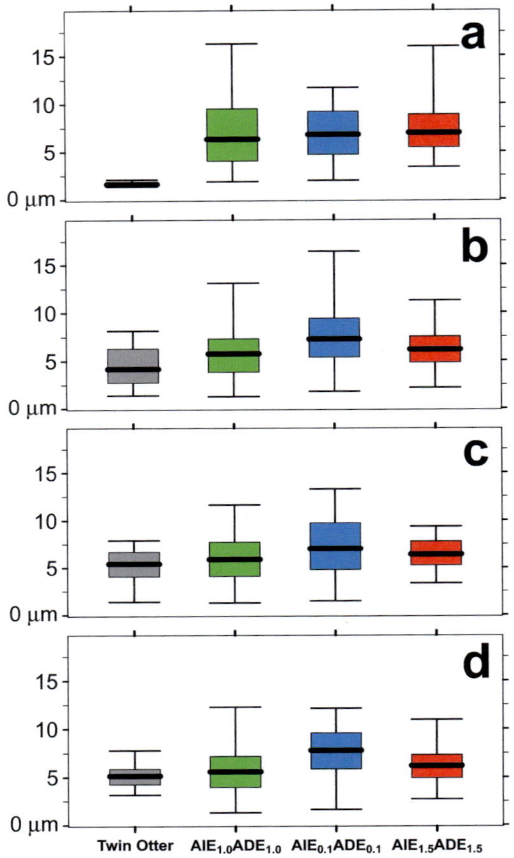

Figure 6.47: Same as Figure 6.46 but for the cloud droplet effective radius (μm) by consistently omitting areas with CDNC below 10 cm^{-3}.

The general problem of separating the effects from the diurnal evolution and the change in the location within the interpretation of aircraft observations is also obvious in this case. The observed cloud properties show no distinct difference between the coastal area and the area in the vicinity of Savè.

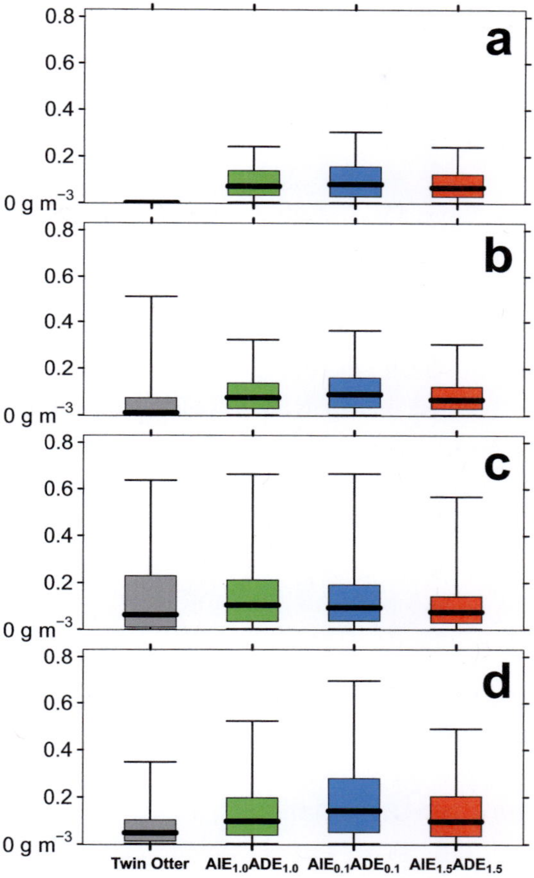

Figure 6.48: Same as Figure 6.46 but for the cloud water (g m^{-3}) by consistently omitting areas with CDNC below 10 cm^{-3}.

Generally higher CDNC and lower effective radii are expected in the vicinity of the downwind Accra plume (blue and black sections and rectangles in Fig. 6.45) than in the remote areas (gray and red sections and rectangles in Fig. 6.45). This can neither be confirmed by the observations nor by the model, also due to the above mentioned problem of simultaneous change in space and time. Nevertheless, it can be stated that the red flight track section (13 UTC) is related to the highest values of observed CDNC, effective radii and cloud water, which is in agreement with the results from the reference case except the cloud water.

Since the model data allows to select a certain area and focusing on the temporal evolution above that area, added value can be provided to the aircraft-model comparison presented above. To further investigate the aerosol impact on clouds in the downwind area of Lomé, the cloud characteristics and meteorological properties for the flight track section between Lomé and Savè (blue section and gray section up to Savè in Fig. 6.45a) are analyzed between 7 UTC and the actual overflight between 11 and 12 UTC in the following section. Since the DACCIWA research aircrafts never conducted measurement flights during night or in the early morning (for safety reasons), the model results can provide additional information on the cloud evolution before the actual research flight took place.

6.4.3 Focus on the NLLS Evolution in the Lomé-Savè Transect Area

Aerosol Amount and Composition

In this section the focus is set on the spatiotemporal evolution of the NLLS in the Lomé-Savè transect area between 7 and 12 UTC to assess whether a change in the aerosol amount affects (a) the extension of the NLLS, (b) the stratus-to-cumulus transition and (c) the cloud microphysical properties also with respect to the expected difference between the coastal area close to Lomé and the inland area near Savè that remained uncertain in the previous section. Figure 6.49 shows the total aerosol mass concentration (Fig. 6.49a) and total aerosol number concentration (Fig. 6.49b) for 3 July 11 UTC according to the

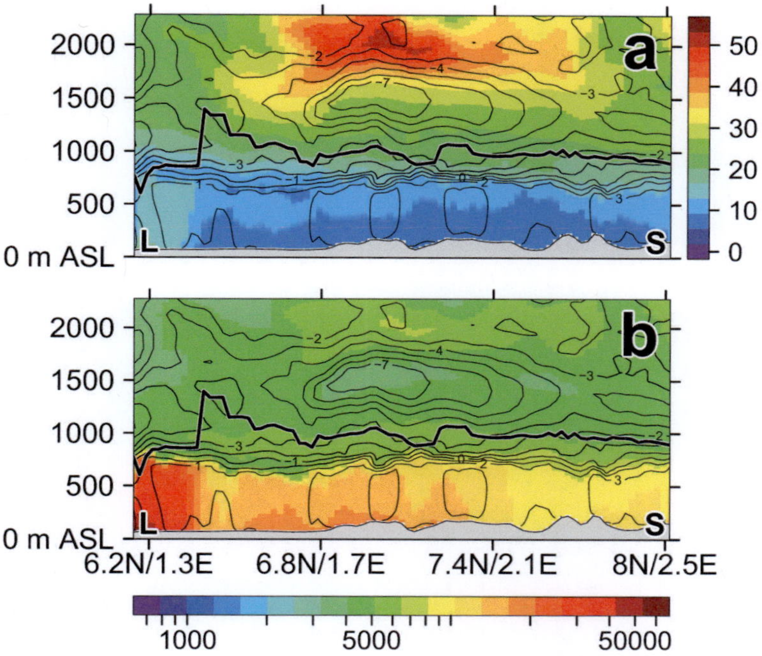

Figure 6.49: Aerosol concentration along the Lomé (L) - Savè (S) vertical transect (m ASL) on 3 July 11 UTC for (a) the total aerosol mass concentration (μg m^{-3}) and (b) the total aerosol number concentration (cm^{-3}) in the rererence case. The contour shows the meridional wind speed (m s^{-1}), the thick black line the monsoon layer depth according to the definition of Fink et al. (2017) and the gray shading denotes the topography.

Twin Otter flight time. The results indicate a low aerosol mass in the lowest 800 m ASL but very high aerosol numbers, especially in the vicinity of Lomé and in general south of the hilly terrain beginning around 6.8°N. Above the monsoon layer (thick black line) the aerosol mass is elevated but the aerosol number is reduced. Figure 6.49 therefore reveals the dominance of small particles (Aitken and accumulation) with negligible mass in the lowest 800 m ASL that is related to the strongest southerly winds.

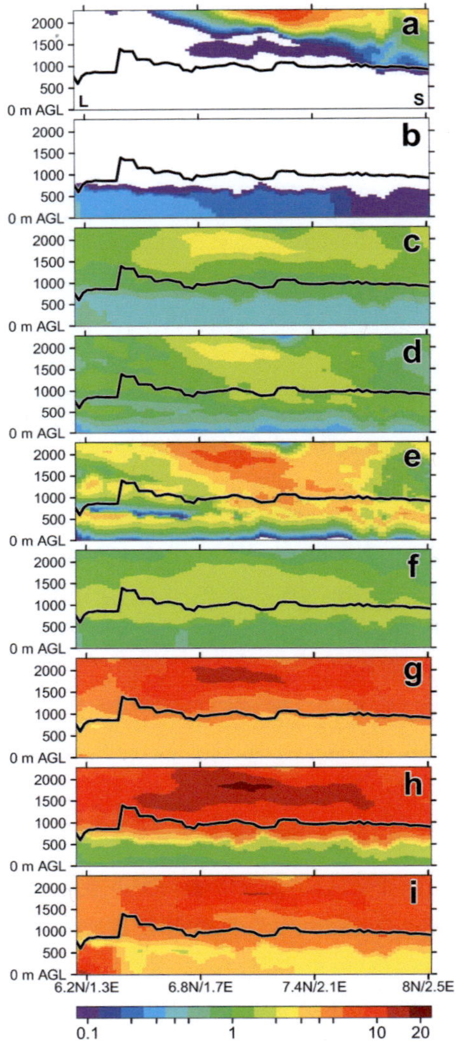

Figure 6.50: Aerosol mass composition (μg m^{-3}) along the Lomé (L) - Savè (S) vertical transect (m AGL) on 3 July 11 UTC for (a) mineral dust, (b) sea salt, (c) black carbon, (d) ammonium, (e) nitrate, (f) sulfate, (g) primary organics, (h) secondary organics and (i) elemental carbon. The black line denotes the monsoon layer depth.

Above the monsoon layer with its easterly winds, larger particles with a larger mass become more important. Figure 6.50 separates the aerosol mass according to its composition. As expected, mineral dust is located above the monsoon layer (Fig. 6.50a) and sea salt within the monsoon layer with higher values near the coast (Fig. 6.50b). The temporal evolution of the atmospheric composition between 23 UTC and 11 UTC (not shown) exhibits a cleaning of the lowest 1 km AGL via the NLLJ. Therefore the larger particles are advected northwards. This is especially remarkable for black carbon (Fig. 6.50c). From the visual inspection of the aerosol advection in the coarse D2 domain, the higher loading of primary and secondary organics above the monsoon layer (Fig. 6.50g-h) result from a northeastward transport from the coastal areas within the monsoon layer, a subsequent upward transport north of 12°N via convective processes and a return transport via northeasterly wind back over SWA (not shown).

NLLS Evolution and Stratus-To-Cumulus Transition

Via Figure 6.51 and 6.52 the full temporal evolution of the NLLS between 2 July 23 UTC and 3 July 12 UTC is assessed. This time period spans the time from the beginning of the inland propagation and intensification of coastal NLLS at 2 UTC to the Twin Otter research flight within the Lomé-Savè transect between 11 and 12 UTC. The left hand side figures show the cloud water for the reference case and the right hand side for the clean case. The 298 K isentrope for the reference (black solid) and the clean case (black dashed) are added in all figures to allow for comparison. The location of the 298 K isentropes clearly shows ongoing advective cooling between 23 UTC and 2 UTC but no significant difference between the reference and the clean case. At least for this transect no meridional shift of the isentrope is visible as observed for the EMFE front in the afternoon but when focusing on the entire SWA domain, indeed a spatial shift can be observed as shown in Figure 6.44.

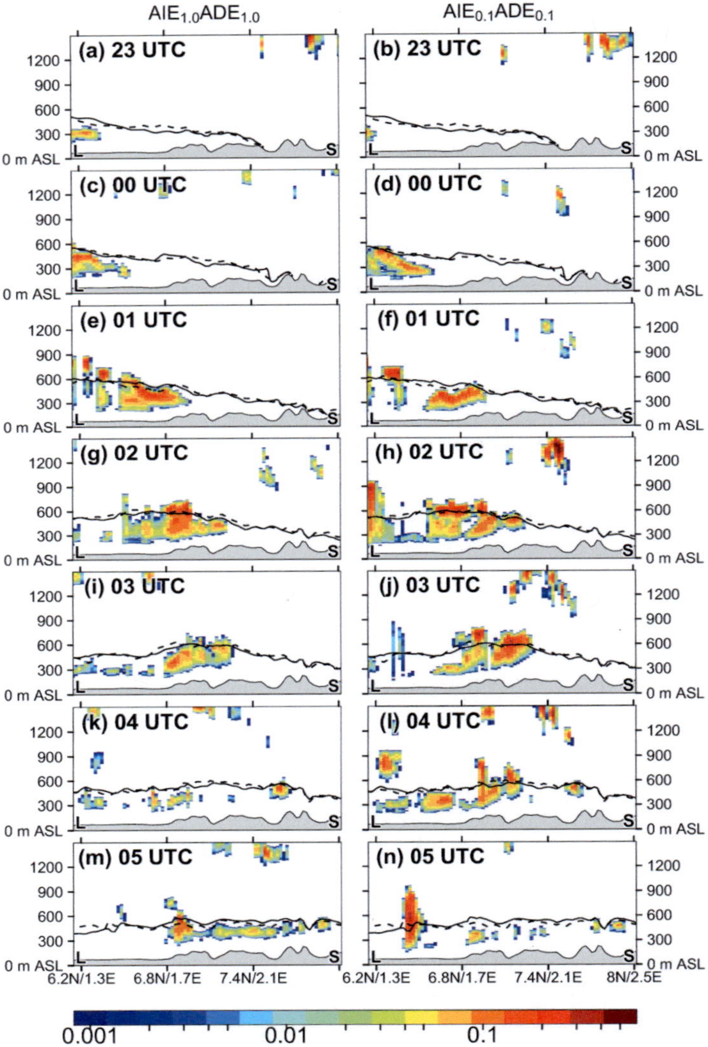

Figure 6.51: Cloud water (g kg^{-1}, shading) along the Lomé (L) - Savè (S) vertical transect (m ASL) for the temporal evolution from 2 July 23 UTC to 3 July 5 UTC (a-n), considering the reference case (left) and the clean case (right). The black solid (dashed) line denotes the 298 K isentrope of the reference (clean) case. To allow for comparison, both isolines are added on the left and right hand side.

After the passage of the 298 K isentrope, NLLS penetrates inland. This first NLLS over land is clearly related to clouds that are formed over the Gulf of Guinea and advected onshore as described in Section 6.4.1. Nevertheless, it can be seen that over land the NLLS intensifies and forms a zonal band that propagates northward (Fig. 6.38). The cloud base between 23 and 5 UTC in Figure 6.51 is relatively homogeneous around 150 m AGL and the vertical extension, primarily limited by the 298 K isentrope, is 300-500 m. The most intense NLLS in terms of vertical extension and cloud water can be found at 2 UTC (Fig. 6.51g,h), covering virtually the complete area south of 7°N. No stringent differences in the progress of the clouds and the cloud coverage are visible between the reference and the clean case.

Figure 6.52 shows the second half of the time period from 6 to 12 UTC (Fig. 6.52a-n). Between 4 UTC (Fig. 6.51k,l) and 8 UTC (Fig. 6.52e,f) the clouds have progressed north of 6.8°N over the hilly area, whereas the area south of 6.8°N is rather cloud free. Sunrise is around 5:30 UTC and at 7 UTC (Fig. 6.52c,d) the increase of the PBL height can be observed. In Figure 6.52 this is illustrated by the area that is related to instability ($d\theta/dz<0$, red shading). For the red shading of the clean case (right hand side of Fig. 6.52) the height of the unstable layer of the reference case is marked by the red line to enable a comparison. From 7 UTC (Fig. 6.52c,d) to 12 UTC (Fig. 6.52m,n) the convective PBL rises in both cases, but generally faster for the clean case than for the reference case. For the area south of 7°N the differences are most significant, in particular between 9 UTC (Fig. 6.52g,h) and 10 UTC (Fig. 6.52i,j), denoting the time of the modeled NLLS maximum in SWA (Fig. 6.38). Since the NLLS layer is lifted with the development of the convective PBL, their different vertical extension in the reference and clean case also leads to a difference in the NLLS cloud base. At 7 UTC (Fig. 6.52c,d) the cloud base is around 200 m AGL for both cases. At 10 UTC (Fig. 6.52i,j) the cloud base south of 6.8°N is around 450 m AGL for the reference case and above 600 m AGL for the clean case with differences on larger scales of up to 300 m.

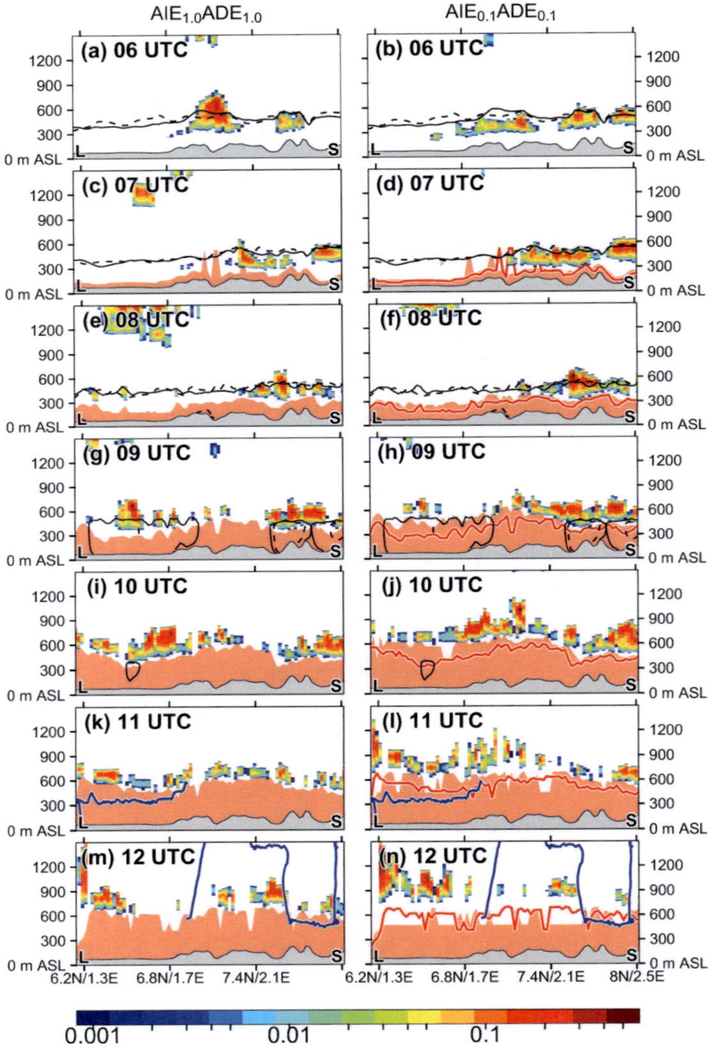

Figure 6.52: Same as Figure 6.51 but for 3 July 6-12 UTC (a-n). The red shading highlights instability ($d\theta/dz<0$). Furthermore, the red line on the right hand side denotes the height of the unstable layer in the reference case to emphasize the differences between the reference and the clean case. The blue lines show the Twin Otter flight track.

At 9 UTC (Fig. 6.52g,h) and 10 UTC (Fig. 6.52i,j) a virtually closed cloud deck develops, also capturing the southern part of the transect that was cloud free in the hours before. This maximum in cloud coverage agrees well with the overall SWA cloud pattern in Figure 6.38. After 10 UTC (Fig. 6.52i,j) the cloud layer tends to break up, highlighting the ongoing stratus-to-cumulus transition. The cloud base and vertical extension of the clouds at 11 UTC (Fig. 6.52k,l) is higher in the clean case than in the reference case. For 11 UTC (Fig. 6.52k,l) and 12 UTC (Fig. 6.52m,n) also the Twin Otter flight track section, related to this time period, is added in blue. 11 UTC (Fig. 6.52k,l) shows the takeoff at Lomé. As mentioned above, the aircraft stays below the cloud in the first part of the flight owing to the minimum safety altitude issue. This allows to compare the observed cloud base in the vicinity of Lomé (300-400 m AGL) with the modeled cloud base in the reference (500-600 m AGL) and clean case (600-700 m AGL). The modeled cloudbase is therefore approximately twice as high as the observed cloud base. For the first Twin Otter cloud profiling between 11:30 UTC and 12 UTC within the Lomé-Savè transect (blue line in Fig. 6.51k-n at 11 and 12 UTC) a cloud base, cloud top and cloud-layer thickness of 730 m AGL, 1200-1300 m AGL and 500 m, respectively, was observed. When comparing this to the modeled situation at 11 UTC (Fig. 6.52k,l), a better agreement can be found for the clean case in which also the observed multiple cloud top heights are traceable.

The study of the NLLS temporal evolution in the first half of 3 July in Figure 6.51 and 6.52 reveals the strongest aerosol impact at 9 and 10 UTC (Fig. 6.52g-j) related to the difference in the cloud base height that is stronger in the southern part of the transect than in the northern part. This southern area is more affected by the Lomé aerosol (Fig. 6.49). It is suggested that the reduced aerosol in the clean case leads to an increase in surface net downward shortwave radiation and therefore to an increased surface heating compared to the reference case, as observed for the study of the EMFE front in Section 6.3. The surplus in heating furthermore intensifies the daytime development of the convective PBL that leads to the differences in the cloud base height.

This is especially interesting since the strongest difference in the cloud base height coincide with the time of the maximum cloud coverage across SWA at 9-10 UTC (compare Fig. 6.38). In addition, the results indicate that the ADE seems to have a much stronger impact than the AIE, as already observed for the EMFE front.

Cloud Properties and Meteorological Quantities During the NLLS Maximum

To come to more robust conclusions, the following figures will focus on the time period 7-10 UTC, capturing the onset of the PBL height increase at 7 UTC and the cloud cover maximum between 9 and 10 UTC. Figure 6.53 shows the Lomé-Savè transect of the median CDNC regarding the vertical column up to 2.3 km AGL (according to the vertical range of the Twin Otter observations, Fig. 6.45b) for 7 UTC, 8 UTC, 9 UTC and 10 UTC (Fig. 6.53 a-d, respectively). It has to be considered that in the following transect figures the ordinate is not the height any more but the range of the meteorological quantity. The black solid (blue dashed) lines denote the reference case (clean case). Furthermore, the existence of clouds along the transect is indicated by the horizontal black (blue) line for the reference case (clean case). A cloud is identified if the total cloud water exceeds 1 g m^{-2}.

As expected, with higher aerosol the CDNC is generally higher if both cases show clouds at the same location. At 7 UTC (Fig. 6.53a) the NLLS is primarily located over the hilly area north of 7.1°N and shows similar CDNC for the reference and clean case. At 8 UTC the NLLS is still present in the northern part with less CDNC differences. In the southern part the clouds are related to higher altitudes (cloud base around 1200 m ASL, Fig. 6.52). This altitude is linked with lower aerosol number concentrations when comparing with Fig. 6.49 and the clouds show stronger CDNC differences than the NLLS in the northern part.

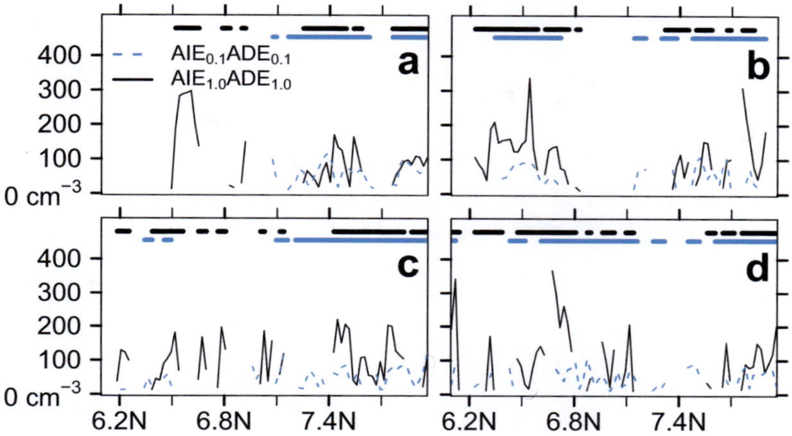

Figure 6.53: Median CDNC (cm^{-3}) for the vertical column up to 2.3 km AGL along the Lomé-Savè transect for 3 July (a) 7 UTC, (b) 8 UTC, (c) 9 UTC and (d) 10 UTC. The black solid (blue dashed) lines denotes the reference case (clean case). Furthermore, the horizontal black (blue) lines on top highlight areas of the vertical transects in the reference case (clean case) that are affected by clouds (total cloud water exceeds 1 g m^{-2}).

At 9 and 10 UTC the clouds are located betwen 450-900 m AGL. The CDNC difference between the reference and the clean case is relatively homogeneous along the transect and also within the realizations no distinct separation between the higher polluted South and the less polluted North is detectable. Therefore also this detailed analysis in addition to Section 6.4.2 fails to identify significant differences. The effective radii (Fig. 6.54) are generally smaller for the reference case but the spatial distribution along the transect and the temporal evolution show no significant differences.

With respect to the feedback of ADE on the lifting of the cloud layer, Figure 6.55 shows the surface net downward shortwave radiation. South of 6.9°N the ADE leads to an increase in the surface shortwave radiation of about 50 W m^{-2} (7 UTC, Fig. 6.55a), 100 W m^{-2} (8 UTC, Fig. 6.55b) and 200 W m^{-2} (9 UTC, Fig. 6.55c) in the clean case compared to the reference case.

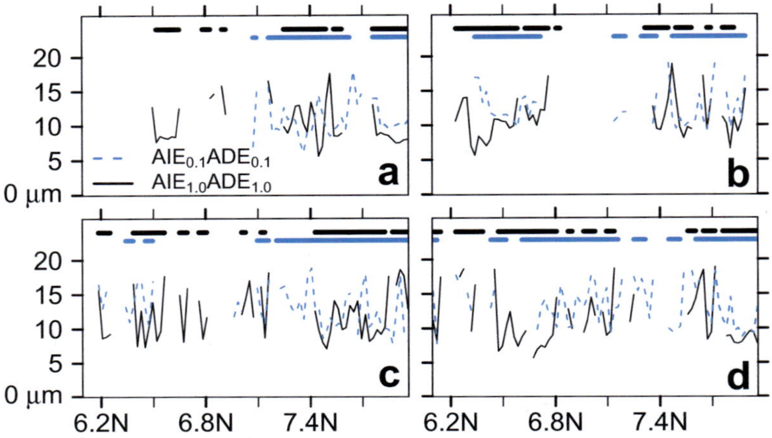

Figure 6.54: Same as Figure 6.53 but for the median cloud droplet effective radius (μm) for the vertical column up to 2.3 km AGL.

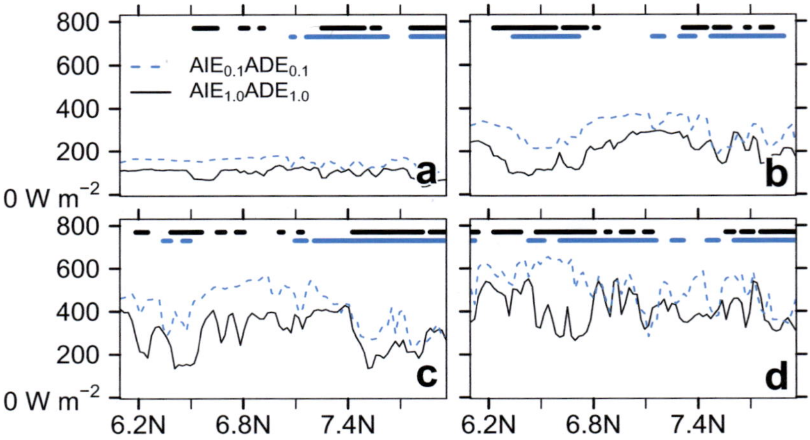

Figure 6.55: Same as Figure 6.53 but for net surface downward shortwave radiation (W m^{-2}).

The latter is comparable with the reduction in surface net downward shortwave radiation due to the NLLS north of 7.4°N. This is a remarkable finding: At 9 UTC, around the time of the maximum NLLS extension, the loss of surface downward shortwave radiation due to the NLLS is similar to the loss due to the ADE when increasing the aerosol mass by one order of magnitude from F_{ADE}=0.1 to F_{ADE}=1.0. Between 7-10 UTC the NLLS is more persistent in the Savè area than in the southern part, which contributes to the establishment of stronger differences in the surface net downward shortwave radiation south of 7.1°N. The surplus in surface net downward shortwave radiation leads to an increase in the 2 m temperature of up to 2-2.5 K in the relevant area as shown in Figure 6.56. Between 7 UTC and 10 UTC a distinct temperature peak develops south of 6.2°N, related to the change in the surface characteristics from vegetated areas to the urbanized area of Lomé.

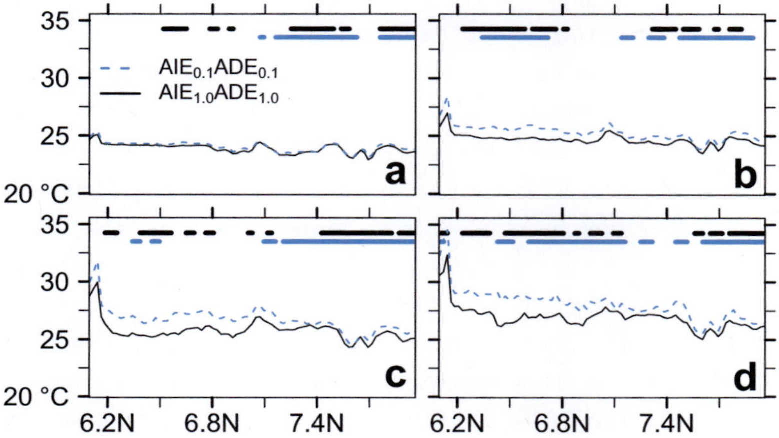

Figure 6.56: Same as Figure 6.53 but for 2 m temperature (°C).

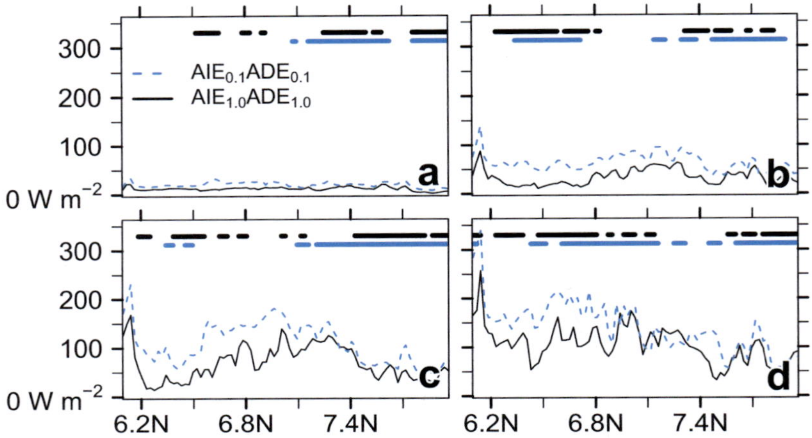

Figure 6.57: Same as Figure 6.53 but for surface sensible heat flux (positive upward, W m^{-2}).

The increased surface heating furthermore leads to an increase in the sensible heat flux of up to 50-100 W m^{-2} in the southern part of the transect for the clean case (Fig. 6.57). These findings support the hypothesis that the aerosol reduction lead, via the ADE, to a stronger surface heating, a faster development of the PBL and finally to a lifting of the clouds to greater heights, which is related to a faster stratus-to-cumulus transition as visible in Figure 6.52 at 11 UTC. As a supplement, Figure 6.58 shows the Lomé-Savè transect at 2 UTC that focuses on the relatively thick (300-450 m) NLLS that has propagated inland from the coast as a zonal band. Since this cloud band is spatially consistent in the clean and reference case, it allows for a more direct comparison. At 2 UTC the coastal NLLS band passed 7.1°N. It is characterized by highest CDNC in front of the cloud band, up to 250 (150) cm^{-3} in the reference case (clean case), and decreasing CDNC behind (Fig. 6.58a). The effective radii are around 10 μm (Fig. 6.58b). When focusing on the net longwave downward radiation (Fig. 6.58c), the area covered by NLLS shows relatively homogeneous values around -10 W m^{-2} in agreement with the NLLS-related observations at Kumasi and Savè supersite

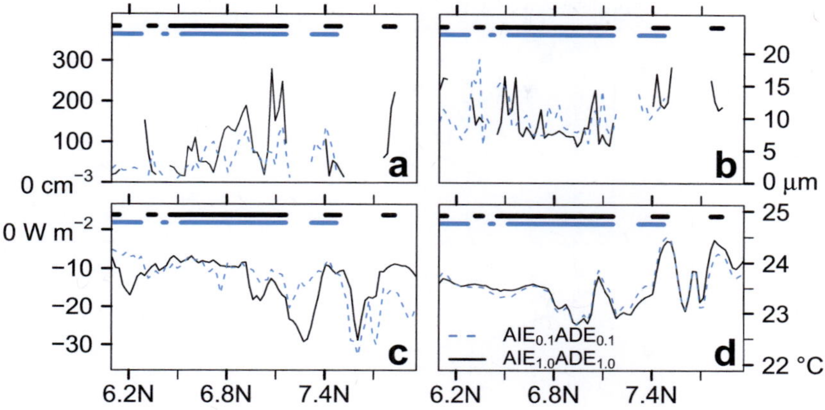

Figure 6.58: Lomé-Savè transect on 3 July 2 UTC for (a) median CDNC (cm^{-3}) for the vertical column up to 2.3 km AGL, (b) median cloud droplet effective radius (μm) for the vertical column up to 2.3 km AGL, (c) net surface downward longwave radiation (W m^{-2}) and (d) 2 m temperature (°C). The black solid (blue dashed) lines denote the reference case (clean case). Furthermore, the horizontal black (blue) lines on top highlight areas of the vertical transects in the reference case (clean case) that are affected by clouds (total cloud water exceeds 1 g m^{-2}).

(Kalthoff et al., 2017). The 2 m surface temperature differences are negligible, which is likely related to the dominance of the cold air advection from the Gulf of Guinea towards the radiative cooling.

Stratus-To-Cumulus Transition Beyond the Lomé-Savè Transect

To assess whether these effects are conferrable to the general SWA conditions, Figure 6.59 shows the temporal evolution of the spatial mean (8°W-3.5°E, 5-10°N) cloud cover (Fig. 6.59a) and cloud base (Fig. 6.59b) for the reference case (black solid) and the clean case (blue dashed).

Between 21 UTC and the time of sunrise (5:30 UTC) the cloud cover increases via clouds that propagate onshore or develop at the topographic barriers. This is linked with a reduction in the mean cloud base.

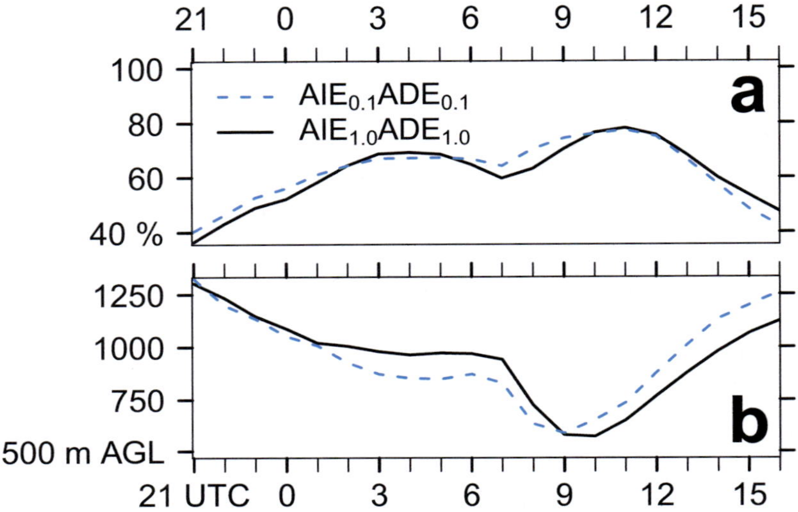

Figure 6.59: Spatial average (8°W-3.5°E, 5-10°N) of (a) total cloud cover (%) and (b) cloud base height (m AGL) for the temporal evolution between 2 July 21 UTC and 3 July 16 UTC. The cloud cover is detected by nonzero values of the total cloud water and the cloud base via the lowest height AGL with a nonzero cloud water value. The black solid (blue dashed) line denotes the reference case (clean case).

Between 1 UTC and 7 UTC the clean case shows lower cloud base values than the reference case. The analysis of the corresponding time reveals that this deviation is not related to NLLS but to mid-level clouds over the Lake Volta Basin and in the northwestern part of the domain. After sunrise, the cloud base decreases via the lifting of near-surface moisture with the initialization of the surface fluxes. The maximum cloud cover is related to the minimum cloud base, underlining the dominance of NLLS. After reaching the cloud cover maximum, the stratus-to-cumulus transition continues, which is related to an increase in the cloud base and a decrease in cloud cover. For this stratus-to-cumulus transition a clear temporal shift of about one hour can be observed between the clean and the reference case. The clean case reacts faster to the insolation after sunrise, reaches the cloud cover maximum earlier and starts earlier with the stratus-to-cumulus transition as observed for the Lomé-Savè transect. This finally leads after 12 UTC to a cloud cover that is reduced compared to the reference case (Fig. 6.59a) implying an additional surplus in net surface shortwave downward radiation that can be used for further heating the surface and accelerating the EMFE front.

Within this section, the NLLS was analyzed by focusing on the spatial extension, the stratus-to-cumulus transition and cloud microphysical properties. The detailed study of the Lomé-Savè transect reveals an inland propagation of coastal stratus between 2 July 23 UTC and approximately 4 UTC. The highest spatial extension is reached between 9 and 10 UTC via the development of the convective PBL and the subsequent lifting of the moist layer above. The stratus-to-cumulus transition is faster in the clean case than in the reference case due to a surplus in net surface downward radiation and therefore in surface heating. As for Section 6.4.2, the analysis of the CDNC and effective radii does not reveal significant differences between the area near to and far from the Lomé city emissions.

The aerosol feedback case study simulations presented in Section 6 revealed several mechanisms relevant for SWA, affecting the location and propagation of the EMFE front and the stratus-to-cumulus transition. In the following section a proposal for a schematic synergy of these mechanisms will be presented.

6.5 A New Conceptual Picture of the Aerosol-Atmosphere Interaction in SWA

This section aims to synthesise the findings that have been received within the aerosol feedback case study. A prevailing meteorological feature of SWA is a front that develops during daytime between the monsoon flow over the ocean and the turbulence over land. This front, being stationary near the coast around noon and propagating inland in the afternoon and evening, was observed in previous studies for several African regions. Since this front propagates in opposite direction as expected for a sea-breeze front, the term *Evening Monsoon flow Enhancement* (EMFE) with an EMFE front and a postfrontal EMFE airmass is proposed in this study. The present study reveals that EMFE affects the entire SWA domain through the course of the day via cold air advection, the NLLJ that can be found in the EMFE postfrontal area, and convergence-induced convection and precipitation. Two distinct meteorological responses to changes in the amount of aerosol via the ADE were identified:

1. A spatial shift of the Evening Monsoon Flow Enhancement (EMFE) and
2. A temporal shift of the Stratus-to-Cumulus Transition (SCT).

Figure 6.60 denotes a conceptual scheme that combines both responses. The bigger circle is related to the first response (EMFE) and the smaller circle to the second response (SCT). The reduction in the amount of aerosol (number and mass) by one order of magnitude is the initial pertubation of the system. The following numbers in parenthesis are related to the spatial average over Ivory Coast on 2 July 15 UTC to provide guiding values for the denoted changes. Via ADE the aerosol reduction leads to an increase in surface net downward shortwave radiation (+70 W m^{-2}) and surface temperature (+0.6 K). Previous studies showed that till the early afternoon, the EMFE front is located stationary near the coast via the balance between the monsoon flow from the sea and the sensible heat flux (turbulence) over land. With the afternoon decrease in sensible heat flux, the EMFE front propagates inland. This study showed that the increased surface heating lead to a negative pressure anomaly over land (-0.26 hPa) and with that to a higher land-sea pressure gradient. The latter is more persistent than the sensible heat flux that vanishes around sunset.

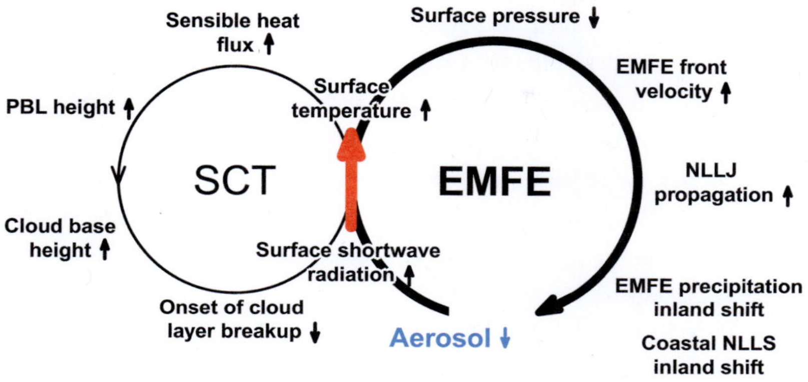

Figure 6.60: Scheme of the aerosol-related atmospheric feedbacks summarizing the findings of the case study simulations on 2-3 July 2016. The main cycle is labeled EMFE (Evening Monsoon Flow Enhancement) and the additional cycle SCT (Stratus-to-Cumulus Transition). The small arrows in upward and downward direction denote whether a quantity reacts with a decrease (downward) or increase (upward) on the decrease of the aerosol mass and number (blue) as the initial pertubation. The red arrow shows the linkage between EMFE and SCT via the increase in shortwave radiation and surface temperature and a potential pathway for a positive feedback of SCT on EMFE.

The stronger pressure gradient leads to a higher EMFE frontal velocity and therefore to a meridional forward shift in the case of reduced aerosol (up to 25 km on 2 July 22 UTC). The postfrontal area is characterized by stronger wind speeds in the lowest 1000 m AGL with the maximum around 250 m AGL that is characteristic for the NLLJ. Therefore an EMFE frontal shift leads to a shift in the NLLJ inland propagation. Since the EMFE frontal propagation is linked to convergence-induced convection and convective precipitation, also a northward shift of the EMFE-related precipitation is observed. These effects are primarily related to the afternoon but the EMFE frontal and NLLJ shift also leads to a shift in the inland propagation of coastal NLLS with a similar spatial magnitude as observed for the EMFE front.

The EMFE cycle denoted in Figure 6.60 intrinsically includes a further mechanisms, related to the counteracting effects of the monsoon flow over the ocean and the sensible heat flux over land in the stationary phase of the EMFE front. With decreasing aerosol the inland sensible heat flux increases, which relocates the front nearer to the coast. Therefore with less aerosol the EMFE frontal inland propagation starts nearer to the coast but is nevertheless faster than in the high aerosol case as soon as the higher pressure gradient dominates towards the decreasing turbulence.

The SCT cycle is coupled to the EMFE cycle via the increase in surface shortwave radiation and temperature. This study pointed out that the surplus in surface heating lead to an increase in sensible heat flux and therefore to an intensified development of the convective PBL. The increase in the PBL height leads to higher cloud bases and therefore to an earlier stratus-to-cumulus transition related to an earlier breakup of the closed cloud layer to scattered cumuli. This is a pathway for a further surplus in surface shortwave radiation and surface heating that could invigorate the EMFE cycle as emphasized by the red arrow in Figure 6.60. However, the latter coupling between the two cycles is only hypothesized. A future study has to assess the significance of the contribution in inland pressure decrease that comes from the surplus in shortwave heating via the earlier cloud-layer breakup. Both cycles are initialized after sunrise with the input in shortwave radiation. The SCT cycle considering less aerosol reveals a negative cloud cover anomaly after 12 UTC. On the other hand the sunset is around 18:30 UTC. Although the EMFE front already starts to penetrate inland around 14-15 UTC, approximately a 6.5-hour period is available for an additional surface heating from the earlier cloud-layer breakup.

The mechanisms described in Figure 6.60 raise the question about the possibility to generalize these results. The EMFE feature is very likely a regular phenomenon under undisturbed monsoon conditions as confirmed by previous studies that focus on longer time periods. Within this case study the EMFE frontal shift was obvious for both days in the evening. However, the results presented above are related to Ivory Coast that shows a more coherent EMFE frontal pattern than the eastern part of the domain, likely related to the topographic features.

This conceptual picture is based on the implications from ADE and does not explicitly include effects from AIE. The AIE assessment within the case study reveals the known mechanisms, in particular the increase (decrease) of the CDNC (effective radius) with an increase in the aerosol number concentration, but has not identified significant differences in cloud microphysical properties with respect to the distance to coastal cities as hotspots of atmospheric pollutant emissions. Also the EMFE-related clouds and precipitation reveal, aside from a meridional shift, no statistically significant difference. In the introductory part to the aerosol feedback case study analysis it was hypothesized, that the strong response of EMFE on changing aerosol likely dominate all subsequent aerosol-induced changes in SWA. With respect to the conceptual picture presented in Figure 6.60 this hypothesis can be confirmed. However, this finding does not exclude the possibility for substantial effects from AIE but a conceptual view as presented for ADE has to be left for subsequent studies.

7 Summary and Conclusions

This study focused on southern West Africa (SWA) to quantify the atmospheric composition and to assess the implications of aerosol on the atmospheric conditions via numerical simulations of the regional-scale model COSMO-ART within the framework of the project Dynamics-Aerosol-Chemistry-Cloud Interactions in West Africa (DACCIWA).

The application of COSMO-ART to the SWA conditions was related to significant further developments of the model that were realized within this study. Significant progress has been reached via the development and implementation of a new mineral dust emission scheme that allows for a globally consistent application and shows good agreement with the results of operational mineral dust simulations. Furthermore, a state-of-the-art emission scheme of biogenic compounds was implemented. Both developments also expand into the COSMO-ART succession model ICON-ART linked with two scientific publications. The atmospheric pollution from gas flaring turned out to be of relevance for SWA but were only considered rudimentarily in the available anthropogenic emission inventories. To overcome these shortcomings, a gas flaring emission parameterization was developed, published and applied to SWA within COSMO-ART. This method for the first time allows to directly calculate the combustion species for single flare spots in high temporal resolution without using proxies (e.g. emission factors).

Based on the numerical simulations and by using the comprehensive observational data obtained during the DACCIWA measurement campaign as an evaluation basis, the following findings are made with respect to the main research questions of this study:

Is it feasible to run COSMO-ART in an operational forecast mode for SWA?

The first question is related to the point that COSMO-ART, with full aerosol and chemistry, was not applied operationally before. There were concerns that the computational cost are too high for an operational usage of COSMO-ART. As part of this study, COSMO-ART was applied operationally to forecast the SWA atmospheric composition between 8 May and 31 July 2016 to support the DACCIWA aircraft campaign flight planning. Forecasts with full aerosol and chemistry were initialized daily with a lead time of 57 hours (aerosol feedbacks were not considered). It turned out that COSMO-ART is applicable as an operational model with reasonable computing resources. Comprehensive scripts were developed to visualize the results that are made available for the forecast users at dacciwa.sedoo.fr. Within this study, the COSMO-ART forecast data were used for an extensive evaluation against observational and model datasets and to derive model climatologies. The evaluation, especially with the Savè supersite data, reveals a good representation of the spatiotemporal characteristics of the nocturnal low-level stratus (NLLS) including the spatial distribution, diurnal cycle, cloud base and nocturnal low-level jet (NLLJ). However, shortcomings in the representation of the daytime temperature maximum and precipitation are detected. The convection parameterization is identified as the major issue, leading to cloud cover overestimations that are linked with a low bias in shortwave radiation and precipitation forecasts with a low discrimination that overestimate the precipitation amount. For future forecast campaigns it should be considered that the studies of Pearson et al. (2013) and Marsham et al. (2013) together with subsequent tests with COSMO-ART indicate the benefit of using explicit convection even with relatively coarse grid mesh sizes.

The natural aerosol is well represented in COSMO-ART with similar performances as the operational chemistry forecast model CAMS. With respect to the anthropogenic trace gas and aerosol concentrations, the evaluation indicates overestimations in COSMO-ART. Further research is necessary to develop an anthropogenic emission dataset that better represents the SWA conditions. The COSMO-ART forecasts are archived and available on demand.

Which typical spatiotemporal patterns of the atmospheric composition can be identified?

From the long-term forecast perspective, less day-to-day variability is observable for the advection of pollutants. The pollution plumes from the coastal cities were transported in northeasterly direction with the monsoon flow, well detectable over long distances (e.g. 220 km inland for Abidjan on 3 July). As identified by previous studies (e.g. Mari et al., 2008), the COSMO-ART results show a long-range transport of biomass burning pollutants from central Africa in agreement with remote sensing observations. Also the emissions from gas flaring are non-negligible in the overall picture of pollutant emissions, especially in coastal areas, and can be transported over long distances. A clear separation between the ocean areas with less total aerosol number concentrations and the land areas with high aerosol number concentrations is visible. The NLLJ acts as a nighttime cleaning mechanism that advects less polluted marine air onshore and pushes the polluted air northwards. This hypothesis is supported by the dominance of aged soot particles in the vicinity of Savè and rather low concentrations of freshly emitted soot.
The anthropogenic aerosol dominates over the natural aerosol. The mineral dust AOD in the DACCIWA campaign period is generally low with values around 0.1 and virtually always below 0.3. The mineral dust influence is generally related to the northern part of the domain and above the monsoon layer. Sea salt is primarily restricted to the monsoon layer whereas the anthropogenic pollutants can be found in all layers. Of particular importance

is the finding that a considerable fraction of the SWA aerosol is water, which is most pronounced in the monsoon layer with its high relative humidity. The aerosol liquid water significantly alters the AOD and with that the feedback on radiation. The mixed accumulation mode contributes the largest water mass to the total aerosol liquid water content. This has to be taken into consideration for the interpretation of airborne aerosol measurements that are primarily related to the analysis of dry aerosol. The aerosol growth due to the water uptake is assessed by previous studies for other regions (e.g. Bian et al., 2014, for the North China Plain) but SWA is virtually unexplored in this regard. The present study therefore highlights the need for further research in SWA within the triangle of *high aerosol burden - high moisture - aerosol radiation interaction*. The results imply an west-east gradient in aerosol burden over SWA. This is likely related to the imbalance of the meridional distribution of coastal cities. Accra, Lomé and Lagos release their pollution plumes in the eastern part of the domain that is enveloped by topographic features in addition. In contrast, the western area, in particular Ivory Coast, is larger and the emissions are primarily related to Abidjan.

Can robust ADE and AIE on SWA clouds and atmospheric dynamics be detected?

The results of a detailed modeling case study of the time period 2-3 July 2016, representing days with exceptional intense NLLS, revealed the impact of a large-scale front over SWA. This front develops during daytime between the monsoon flow over the ocean and the turbulence over land being stationary near the coast around noon and propagating inland in the afternoon and evening. Although this phenomenon is already identified for several African coastal regions, including Mauritania (Grams et al., 2010) and SWA (Adler et al., 2017) and conceptually separated from the classical land-sea breeze, it is not termed uniformly. This study proposed the term *Evening Monsoon Flow Enhancement* (EMFE) that is linked with an EMFE front and a postfrontal

EMFE airmass. The EMFE postfrontal area is characterized by a distinct decrease in temperature and increase in wind speed and relative humidity but not specific humidity in agreement with the findings of Adler et al. (2017). Furthermore it turned out that, at least in the case study period, the NLLJ is a EMFE-postfrontal phenomenon on SWA spatial scale rather than a locally induced feature. The arrival of the EMFE front denotes the onset of the NLLJ and the advection of cold air. The inland propagation of the EMFE front is related to convergence-induced convection and precipitation.

An aerosol sensitivity study is applied via the scaling of the aerosol number and mass by the factors 0.1, 1.0 and 1.5, denoted as clean, reference and polluted, respectively. The modeling study reveals a strong aerosol susceptibility of the EMFE frontal location that has not been documented before. With decreasing aerosol the EMFE front shows higher propagation velocities and therefore a meridional inland shift with respect to the realizations with higher aerosol burden, primarily related to the ADE. Grant and van den Heever (2014) modeled a similar behavior for the sea breeze over Cameroon.

With less aerosol, the surplus in incoming solar radiation leads to an intensification in the near-suface heating and therefore to a negative surface-pressure anomaly. The increased land-sea pressure gradient accelerates the EMFE front, leading to an inland displacement of the NLLJ, EMFE precipitation and the coastal NLLS of up to 25 km. The sensible heat flux, as an obstacle for the EMFE frontal inland propagation, diminishes rapidly around sunset whereas the negative pressure anomaly persists longer allowing for the EMFE frontal acceleration. Intensified longwave-cooling with less aerosol might influence the EMFE prefrontal area but even after sunset the positive temperature anomaly from daytime solar heating persists and dominates.

Additionally, the surplus in near-surface heating leads to an earlier stratus-to-cumulus transition via an earlier onset of the convective boundary layer development. The latter results in an earlier lifting of the cloud base. With the earlier stratus-to-cumulus transition the stratus layer breakup occurs earlier, likely further increasing the incoming shortwave radiation compared to the realization with high aerosol burden. It is hypothesized that this radiation surplus is available for further increasing the EMFE meridional shift. The present study synergized this subtle aerosol-atmosphere feedback in a new conceptual scheme combining the EMFE cycle and stratus-to-cumulus cycle (Fig. 6.60). Saleeby et al. (2014) identified AIE as relevant for the stratus-to-cumulus transition over tropical oceans with an accelerated transition with increasing aerosol. This study identified the ADE as the predominant effect for the stratus-to-cumulus transition over tropical land areas with an accelerated transition with decreasing aerosol. For Northern India, Lau et al. (2017) revealed that ADE dominates the large-scale aerosol-monsoon interactions. The present study transfers this outcome also to SWA. In the current aerosol research, ADE is underrepresented towards AIE. Especially with respect to the monsoon, a special focus has to be set on ADE.

The AIE shows less distinct signals in SWA than ADE. Although the known mechanisms, in particular the increase (decrease) of the cloud droplet number concentration (effective radius) with an increase in the aerosol number concentration, can be observed, no coherent SWA-specific AIE features were identified. The ADE-induced meridional shift of the EMFE front also leads to a shift of convective clouds and precipitation in the afternoon and the coastal NLLS during night. Furthermore, the COSMO-ART simulations with increased aerosol number concentrations show tendencies for more localized precipitation with higher precipitation amounts in these local areas, likely due to convective invigoration (Rosenfeld et al., 2008). This precipitation localization supports the findings of Grant and van den Heever (2014).

In which respect are low-level clouds affected by changes in the aerosol amount?

The question about the aerosol effect on low-level clouds accompanies the previous question by highlighting the results for the NLLS that is identified by many studies (e.g. Schrage and Fink, 2012; Schuster et al., 2013; Adler et al., 2017) and also in the present study as the most prevailing cloud feature in SWA during the WAM. Based on the COSMO-ART high resolution case-study simulation of the time period 2-3 July 2016, highest NLLS coverage is modeled at or upstream of the topographic features (as modeled by Adler et al., 2017), especially at the Mampong Range with total cloud water values around 1 kg m^{-2}. Additionally, NLLS over the Gulf of Guinea penetrates inland after 21 UTC, revealing a closed zonal cloud band in particular east of the Atakora Mountains with a cloud base around 150 m AGL and a thickness of 300-500 m. The aerosol-induced meridional shift of the EMFE front shifts this cloud band inland in case of reduced aerosol. The SWA maximum cloud cover is reached around 9-10 UTC in agreement with the findings of Schuster et al. (2013), dominated by NLLS that shows a coherent cloud base between 500-750 m AGL and a total cloud water path of around 20 g m^{-2}. A value of around -10 W m^{-2} for the surface net downward longwave radiation in NLLS environments is modeled in agreement with the observations at Savè and Kumasi (Kalthoff et al., 2017). No significant differences in NLLS microphysical properties were modeled in the comparison of areas near to and far from coastal cities.

Especially the findings of the case study analysis can be of great value for the interpretation of the measurements obtained at the DACCIWA supersites by embedding the local results into the larger scale context of SWA. Since the case study is embedded in the monsoon post-onset phase with virtually undisturbed monsoon conditions for nearly one month, it is expected that the findings of the case study can be generalized to this time period. Furthermore, this study comprises the rather exceptional combination of both, model realizations on a monthly time scale with a coarse grid mesh size and model realizations on a daily time scale with a fine grid mesh size and the consideration of aerosol feedbacks, to assess the SWA air pollution and meteorological characteristics.

Several research questions remain unanswered or additional questions arise during the data analysis that could be incorporated in this study. Three points were identified that could lead to significant progress in the understanding of

the results. The first aspect is related to the temporal extension of the feedback simulations and the application of further aerosol factors. An increase in the modeled time period, which is feasible in terms of computational cost, could not only lead to more robust results with respect to the EMFE frontal shift, but might also shed light on the question why the NLLS is more persistent on 3 July than on other days.
The second point is the comprehensive evaluation of the case study simulations with the data obtained from the DACCIWA measurement campaign as it is done for the COSMO-ART forecasts. Furthermore, it would be interesting to see whether the aerosol-related EMFE frontal shift can also be observed in the real world. In modeling, the environment is controlled, such the heat low and monsoon flow is unchanged and only the SWA conditions are altered. In the real world it is more complicated to disentangle the aerosol effect from other variations. One possibility might be to focus on the location of the EMFE front during their stationary phase around 12-14 UTC. In this phase the EMFE front is nearer to the coastline with less aerosol. The EMFE front can be detected via the MSG/Seviri cloud product in high temporal resolution since the area directly behind the EMFE front appears to be cloud free. With a statistical approach, a clustering of days with high and low AODs can be used to identify potential relationships between the aerosol amount and the EMFE frontal distance to the coast. However, the MODIS AOD observation during the WAM is strongly limited due to clouds and it is questionable whether days with reduced AODs can be detected in SWA. During the boreal winter, SWA is affected to a greater extent by local biomass burning likely leading to higher AODs and simultaneously the cloud cover is reduced during that time of the year. Therefore the winter period could be more effective for the EMFE-aerosol sensitivity study by using fire detections of MODIS or VIIRS as an indicator for the aerosol burden over SWA. However, in this time of the year the Harmattan and not the monsoon flow drives the circulation pattern over SWA. One possibility would be to focus on the transition times spring and autumn in which likely both, EMFE and the SWA biomass burning takes place. Also

the historic development in SWA could be used to assess the EMFE-aerosol sensitivity. By considering the positive trend of anthropogenic emissions in SWA, the question can be posed whether the EMFE front location correlates with this positive emission trend. It is hypothesized that the daytime stationary EMFE front location has shifted inland from the past to the current situation.

The far-reaching point of anthropogenic emissions is the most important issue. The representation of the anthropogenic emissions carries the largest uncertainty in the model realizations. This study used a clean, reference and polluted case but it remains unclear whether the reference case is consistent with the actual conditions. There are indications that in the reference case the aerosol amount is already higher than observed. Therefore we see the need to comprehensively consolidate all available aerosol products, especially the new data obtained during the DACCIWA measurement campaign, to get an impression of the actual aerosol state. With that, a redefinition of the reference case can be done to gain further confidence in the results. However, even by considering the aerosol data from the DACCIWA measurement campaign, the overall aerosol database is strongly limited and high uncertainty will persist. In the short term this problem can be addressed by implementing the emission inventory that was developed in the framework of DACCIWA and that also incorporates local observations from SWA. This inventory tends to have more emissions of black carbon and organic carbon and less emissions of carbon monoxide and nitrogen oxide and dioxide in SWA compared to the EDGAR database used in this study. Another way to face this problem is to neglect all anthropogenic emissions and to analyze the difference between this non-anthropogenic state and the current air pollution state to assess the AIE and the spatiotemporal NLLS structure.

In the long term, the need for an operational air pollution monitoring in the coastal cities and surburban areas of SWA is obvious. With continous observations, the main aerosol sources and amounts could be identified and comprehensively used for the adaptions in air pollution modeling. DACCIWA can provide an insight in the air pollution situation and highlight the need for long-term observations but cannot itself provide extensive climatological views based three weeks of aircraft measurements. Even with an extensive ground-based aerosol network, a three-week measurement period could hardly provide a representative database. The local authorities are called to establish an operational air pollution monitoring. This will be the fundamental first step for all subsequent well scientificly based aerosol-cloud interaction studies, including the separation between local and remote pollution (e.g. from biomass burning), and also the inducement for the authorities to reduce the air pollution in the cities to protect the health of their inhabitants.

A Code and Data Availability

The gas flaring emission parameterization, developed in this study, is available at https://www.zenodo.org/ (doi:10.5281/zenodo.61151, Deetz and Vogel, 2016), entitled "Gas flaring emission estimation parameterization v2". In case of a publication using this code, please cite: Deetz, K. and Vogel, B.: Development of a new gas-flaring emission dataset for southern West Africa, Geosci. Model Dev., 10, 1607-1620, https://doi.org/10.5194/gmd-10-1607-2017, 2017. The gas flaring emission inventory (GFEI) is available for the DACCIWA research community via the DACCIWA database (http://baobab.sedoo.fr/Data-Search/?datsId=1712&terms=flaring&allKeywords=1&project_name=BAOBAB, last access: 10 June 2017) or you can contact the author. The COSMO-ART forecasts, conducted between 8 May and 31 July 2016, can be made available on demand.

B Flaring Emission Time Series

As introduced in Section 3.1.3, a gas flaring emission inventory (GFEI) was derived including six inventories (GFEI-1 to GFEI-6) for the 364 flares within SWA. Figure 3.11 has shown the results for CO and the figures in this section present the emission time series for CO_2, NO, NO_2 and SO_2 in addition (Fig. B.1-B.4, respectively).

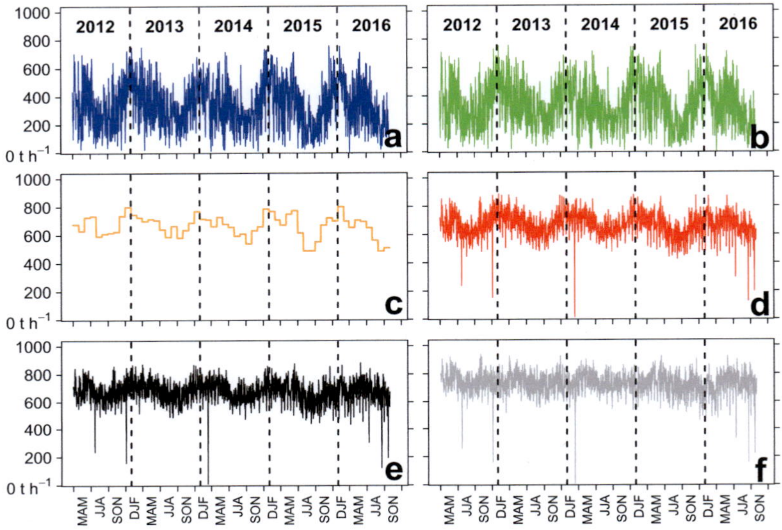

Figure B.1: Gas flaring emission inventory (GFEI) 1-6 for CO_2 (t h^{-1}), spatially integrated over SWA and capturing the time period March 2012 to August 2016. (a) GFEI-1: Using instantaneous (daily) data of source temperature and radiant heat by considering only the directly observed flares, (b) GFEI-2: Using instantaneous (daily) data of radiant heat but monthly mean of source temperature for the respective flare by considering only the directly observed flares, (c) GFEI-3: All 364 flares are active at once using their monthly mean source temperature and radiant heat, (d) GFEI-4: Sum of GFEI-2 and the emissions from cloud-covered flares (not observed but assumed to be active) using the flare specific yearly emission average from GFEI-2, (e) GFEI-5: As GFEI-4 but using the flare specific DJF emission average from GFEI-2 for the cloud-covered flares and (f) GFEI-6: As GFEI-5 but with removal of the annual cycle. The abscissa denotes the seasons and the vertical dashed lines separated the single years.

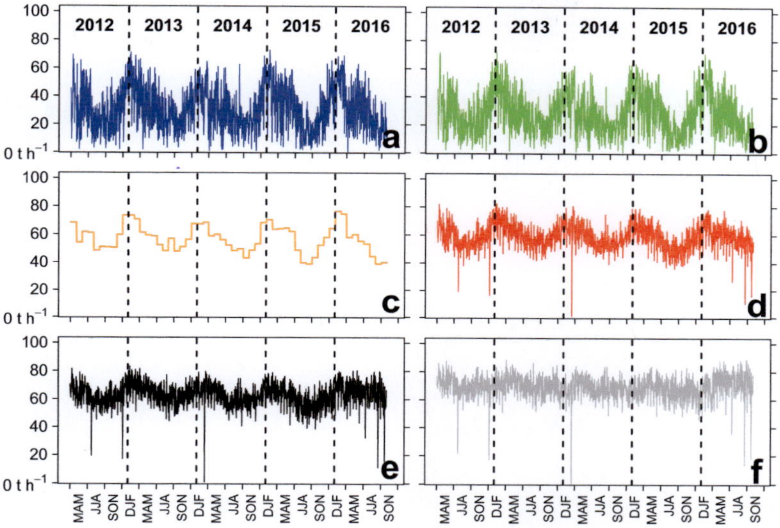

Figure B.2: Same as Figure B.1 but for NO ($t\ h^{-1}$).

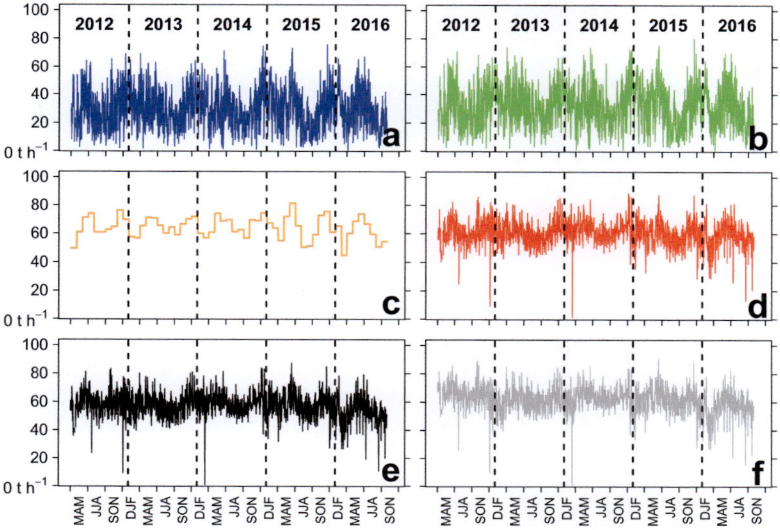

Figure B.3: Same as Figure B.1 but for NO_2 ($t\ h^{-1}$).

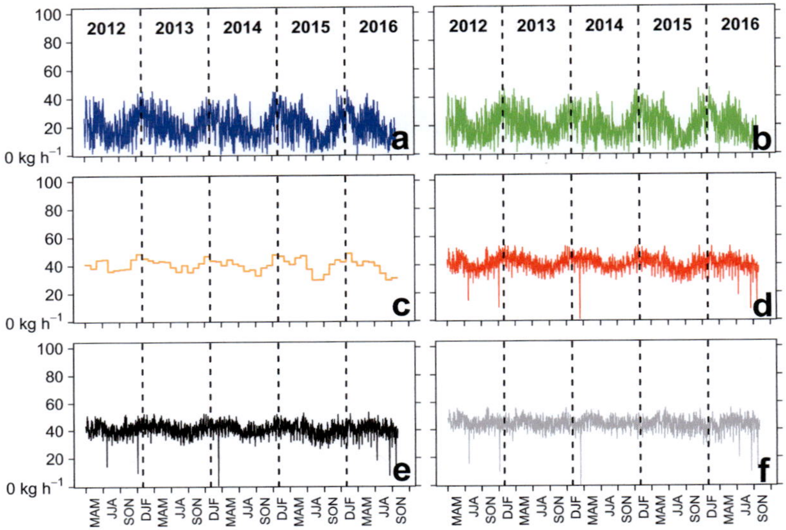

Figure B.4: Same as Figure B.1 but for SO_2 (kg h^{-1}). Notice the different unit of the ordinate compared to the figures of the other species.

C Skill Scores

Four skill scores were used to assess the model performance regarding precipitation: the *Thread Score* (TS with 1=perfect, Eq. C.1), *Probability of False Detection* (POFD with 0=perfect, Eg. C.2), *Peirce Skill Score* (PSS with 1=perfect, Eq. C.3) and *Heidke Skill Score* (HSS with 1=perfect, Eq. C.4). The four entries of the contingency table are abbreviated with (a) hit, (b) false alarm, (c) missed and (d) correct rejected.

$$TS = \frac{a}{a+b+c} \tag{C.1}$$

$$POFD = \frac{b}{b+d} \tag{C.2}$$

$$PSS = \frac{ad-bc}{(a+c)(b+d)} \tag{C.3}$$

$$HSS = \frac{2(ad-bc)}{(a+c)(c+d)+(a+b)(b+d)} \tag{C.4}$$

D Statistical Significance Test of the EMFE Frontal Acceleration

In addition to the significance analysis of the EMFE frontal velocity in Figure 6.19, Figure D.1 shows the analysis for the acceleration of the EMFE front. Figure D.1 reveals, as expected, that significant differences in the mean EMFE front accelerations between the realizations are located in the acceleration phase. After sunset (18:32 UTC) no more significant differences are dectected. It has to be considered that the calculation of the acceleration needs four model time steps (four hours in case of COSMO-ART) and therefore the possibilities of an evaluation are limited.

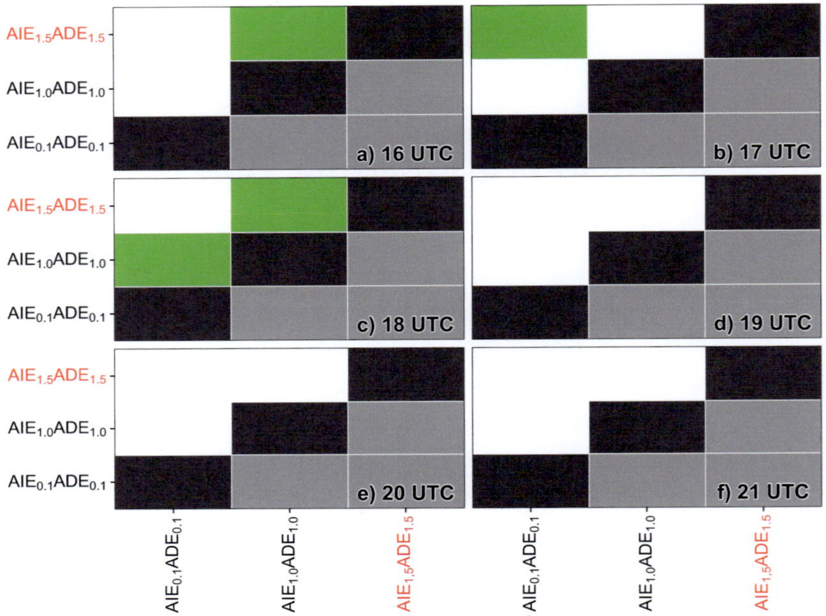

Figure D.1: Statistical significance of the pairwise difference between the experiment averages of EMFE front velocity for 2 July 2016, 16-21 UTC. Green boxes indicate a significant difference whereas white boxes denote no significant difference (Wilcoxon-Mann-Whitney rank sum test with α=0.05). The acceleration at 16 UTC is calculated from the velocity of the front between 15-16 UTC and 16-17 UTC (analog for the other times).

E Factorial Method

Via the factorial method the contributions of AIE and ADE as well as their synergistic effect to the change in SWA meteorology can be isolated. This method is described in detail in Montgomery (2005) and applied in several previous studies (e.g. Teller and Levin, 2008; Kraut, 2015; Rieger et al., 2017). For this study the unreplicated n^k factorial design was used with $n = 2$ describing the number of parameters (AIE and ADE) and $k = 2$ describing the number of different values used for n (0.1 and 1.0). Unreplicated denotes the fact that only one value is available per grid box (no replications). This leads to a 2^2 experiment requiring the first four simulations denoted in Table 6.1. The relative contributions are derived by calculating the sum of squares of AIE (SS_{AIE}), ADE (SS_{ADE}), their synergistic interaction SYN (SS_{SYN}) and the total effect TOT (SS_{TOT}) in Equation E-E.4, respectively:

$$SS_{AIE} = \frac{1}{4}\,(AIE_{1.0}ADE_{1.0} + AIE_{1.0}ADE_{0.1} - AIE_{0.1}ADE_{1.0} - AIE_{0.1}ADE_{0.1})^2, \tag{E.1}$$

$$SS_{ADE} = \frac{1}{4}\,(AIE_{1.0}ADE_{1.0} + AIE_{0.1}ADE_{1.0} - AIE_{1.0}ADE_{0.1} - AIE_{0.1}ADE_{0.1})^2, \tag{E.2}$$

$$SS_{SYN} = \frac{1}{4}\,(AIE_{1.0}ADE_{1.0} - AIE_{1.0}ADE_{0.1}$$

$$- AIE_{0.1}ADE_{1.0} + AIE_{0.1}ADE_{0.1})^2 \text{ and} \tag{E.3}$$

$$SS_{TOT} = SS_{AIE} + SS_{ADE} + SS_{SYN}. \tag{E.4}$$

The contributions of AIE, ADE and SYN to the total change in percent are given by Equation E.5-E.7, respectively:

$$c_{AIE} = \frac{SS_{AIE}}{SS_{TOT}}\,100, \tag{E.5}$$

$$c_{ADE} = \frac{SS_{ADE}}{SS_{TOT}}\,100 \text{ and} \tag{E.6}$$

$$c_{SYN} = \frac{SS_{SYN}}{SS_{TOT}}\,100. \tag{E.7}$$

References

Ackermann, I. J., Hass, H., Memmesheimer, M., Ebel, A., Binkowski, F. S., and Shankar, U.: Modal aerosol dynamics model for Europe: development and first applications, Atmos. Environ., 32, 2981–2999, 1998.

Adler, B., Kalthoff, N., and Gantner, L.: Nocturnal low-level clouds over southern West Africa analysed using high-resolution simulations, Atmos. Chem. Phys., 17, 899–910, 2017.

AERONET: Aerosol Robotic Network (https://aeronet.gsfc.nasa.gov/new_web/data_usage.html, last access: 10 June 2017), 2017.

Aitchison, J. and Brown, J. A. C.: The Lognormal Distribution Function, Cambridge Univ. Press, Cambridge, UK, 1957.

Albrecht, B. A.: Aerosols, Cloud Microphysics, and Fractional Cloudiness, Science, 245, 1227–1230, 1989.

Alfaro, S. C. and Gomes, L.: Modeling mineral aerosol production by wind erosion: Emission intensities and aerosol size distributions in source areas, J. Geophys. Res., 106, 18 075–18 084, 2001.

Altstädter, B., Platis, A., Wehner, B., Scholtz, A., Wildmann, N., Hermann, M., Käthner, R., Baars, H., Bange, J., and Lampert, A.: ALADINA - an unmanned research aircraft for observing vertical and horizontal distributions of ultrafine particles within the atmospheric boundary layer, Atmos. Meas. Tech., 8, 1627–1639, 2015.

Amekudzi, L. K., Yamba, E. I., Preko, K., Asare, E. O., Aryee, J., Baidu, M., and Codjoe, S. N. A.: Variabilities in Rainfall Onset, Cessation and Length of Rainy Season for the Various Agro-Ecological Zones of Ghana), Climate, 3, 416–434, 2015.

API: Pressure-relieving and Depressuring Systems, ANSI/API STANDARD 521 FIFTH EDITION, JANUARY 2007, ISO 23251 (Identical), Petroleum and natural gas industries – Pressure-relieving and depressuring systems, Section 7.3.2.4: Design details h), Atmos. Environ., p. 127, 2007.

Athanasopoulou, E., Vogel, H., Vogel, B., Tsimpidi, A. P., Pandis, S. N., Knote, C., and Fountoukis, C.: Modeling the meteorological and chemical effects of secondary organic aerosols during an EUCAARI campaign, Atmos. Chem. Phys., 13, 625–645, 2013.

Athanasopoulou, E., Rieger, D., Walter, C., Vogel, H., Karali, A., Hatzaki, M., Gerasopoulos, E., Vogel, B., Giannakopoulos, C., Gratsea, M., and Roussos, A.: Fire risk, atmospheric chemistry and radiative forcing assessment of wildfires in eastern Mediterranean, Atmos. Environ., 95, 113–125, 2014.

Baer, M. and Nester, K.: Parametrization of trace gas dry deposition velocities for a regional mesoscale diffusion model, Ann. Geophysicae, 10, 912–923, 1993.

Bahino, J., Yoboué, V., Galy-Lacaux, C., Adon, M., Akpo, A., Keita, S., Liousse, C., Gardrat, E., Chiron, C., Ossohou, M., Gnamien, S., and Djossou, J.: Spatial distribution of gaseous pollutants (NO_2, SO_2, NH_3, HNO_3 and O_3) in Abidjan, Côte d'Ivoire, Atmos. Chem. Phys., submitted, 2017.

Baldauf, M., Seifert, A., Förstner, J., Majewski, D., and Raschendorfer, M.: Operational Convective-Scale Numerical Weather Prediction with the COSMO model: Description and Sensitivities, Mon. Wea. Rev., pp. 3887–3905, 2011.

Bangert, M.: Eine Parametrisierung für die Berechnung von Photolysefrequenzen in Chemie-Transport-Modellen, Institut für Meteorologie und Klimaforschung der Universität Karlsruhe (TH), Seminararbeit, 2007.

Bangert, M.: Interaction of Aerosol, Clouds, and Radiation on the Regional Scale, Institut für Meteorologie und Klimaforschung, Karlsruher Institut für Technologie, Dissertation, 2012.

Bangert, M., Nenes, A., Vogel, B., Vogel, H., Barahona, D., Karydis, V. A., Kumar, P., Kottmeier, C., and Blahak, U.: Saharan dust event impacts on cloud formation and radiation over Western Europe, Atmos. Chem. Phys., 12, 4045–4063, 2012.

Barahona, D. and Nenes, A.: Parameterizing the competition between homogeneous and heterogeneous freezing in cirrus cloud formation - monodisperse ice nuclei, Atmos. Chem. Phys., 9, 369–381, 2009a.

Barahona, D. and Nenes, A.: Parameterizing the competition between homogeneous and heterogeneous freezing in cirrus cloud formation - polydisperse ice nuclei, Atmos. Chem. Phys., 9, 5933–5948, 2009b.

Barahona, D., West, R. E. L., Stier, P., Romakkaniemi, S., Kokkola, H., and Nenes, A.: Comprehensive accounting for the effect of giant CCN in cloud activation parameterizations, Atmos. Chem. Phys., 10, 2467–2473, 2010.

Baró, R., Palacios-Peña, L., Baklanov, A., Balzarini, A., Brunner, D., Forkel, R., Hirtl, M., Honzak, L., Pérez, J. L., Pirovano, G. an San José, R., Schröder, W., Werhahn, J., Wolke, R., Žabkar, R., and Jiménez-Guerrero, P.: Regional effects of atmospheric aerosol on temperature: an evaluation of an ensemble of online coupled models, Atmos. Chem. Phys., 17, 9677–9696, 2017.

Bian, Y. X., Zhao, C. S., Ma, N., Chen, J., and Xu, W. Y.: A study of aerosol liquid water content based on hygroscopicity measurements at high relative humidity in the Northern China Plain, Atmos. Chem. Phys., 14, 6417–6426, 2014.

Binkowski, F. S. and Shankar, U.: The Regional Particulate Matter Model: 1. Model description and preliminary results, J. Geophys. Res., 100, 26 191–26 209, 1995.

Blackadar, A. K.: Boundary layer wind maxima and their significance for the growth of nocturnal inversions, Bull. Amer. Meteor. Soc., 38, 283–290, 1957.

Bott, A.: A Positive Definite Advection Scheme Obtained by Nonlinear Renormalization of the Advective Fluxes, Mon. Wea. Rev., 117, 1006–1015, 1989.

Boucher, O., Randall, D., Artaxo, P., Bretherton, C., Feingold, G., Forster, P., Kerminen, V.-M., Kondo, Y., Liao, H., Lohmann, U., Rasch, P., Satheesh, S., Sherwood, S., B., S., and Zhang, X.: Clouds and Aerosols. In: Climate Change 2013: The Physical Science Basis. Contribution of Working Group I to the Fifth Assessment Report of the Intergovernmental Panel on Climate Change, Cambridge University Press, Cambridge, United Kingdom and New York, NY, USA, pp. 576–594,604–623, 2013.

Brook, R. R.: The Koorin nocturnal low-level jet, Bound.-Layer Meteor., 32, 133–154, 1985.

CAMS: Copernicus Atmosphere Monitoring Service (http://apps.ecmwf.int/datasets/data/cams-gfas/, last access: 30 July 2017), 2017a.

CAMS: Copernicus Atmosphere Monitoring Service (http://apps.ecmwf.int/datasets/data/cams-nrealtime/levtype=sfc/, last access: 10 June 2017), 2017b.

Caniaux, G., Giordani, H., Redelsperger, J.-L., Guichard, F., Key, E., and Wade, M.: Coupling between the Atlantic cold tongue and the West African monsoon in boreal spring and summer, J. Geophys. Res., 116, 2011.

Capes, G., Murphy, J. G., Reeves, C. E., McQuaid, J. B., Hamilton, J. F., Hopkins, J. R., Crosier, J., Williams, P. I., and Coe, H.: Secondary organic aerosol from biogenic VOCs over West Africa during AMMA, Atmos. Chem. Phys., 9, 3841–3850, 2009.

CCSM: https://svn-ccsm-inputdata.cgd.ucar.edu/trunk/inputdata/lnd/clm2/raw data/pftlanduse.3minx3 min.simyr2000.c110913/ (last access: 10 June 2017), 2015.

Chang, C.-P., Ding, Y., Lau, N.-C., Johnson, R. H., Wang, B., and Yasunari, T.: The Global Monsoon System, Chapter 8: The West African Monsoon, World Scientific Series on Asian-Pacific Weather and Climate, 2nd Edition, 5, 111–135, 2011.

Chen, G. and Wang, W.-C.: Aerosol-Stratocumulus-Radiation Interactions over the Southeast Pacific: Implications to the Underlying Air-Sea Coupling, J. Atmos. Sci., 73, 2759–2771, 2016.

CMSAF: Operational Products: CFC – Fractional cloud cover instantaneous data (MSG disk, CM SAF definition), Version 350, https://wui.cmsaf.eu/safira/action/viewPeriodEntry?id=11495_14063_15657_15672_16574_19152_ 20532_21207 (last access: 14 June 2017), 2017.

Cook, K. H.: Generation of the African Easterly Jet and Its Role in Determining West African Precipitation, J. Climate, 12, 1165–1184, 1999.

Dai, Y., Dickinson, R. E., and Wang, Y.-P.: A Two-Big-Leaf model for Canopy Temperature, Photosynthesis, and Stomatal Conductance, J. Climate, 12, 2281–2299, 2004.

Deetz, K. and Vogel, B.: Development of a new gas-flaring emission dataset for southern West Africa, Geosci. Model Dev., 10, 1607–1620, 2017.

Deetz, K., Klose, M., Kirchner, I., and Cubasch, U.: Numerical simulation of a dust event in northeastern Germany with a new dust emission scheme in COSMO-ART, Atmos. Environ., 126, 87–97, 2016.

Deng, Z., Zhao, C., Zhang, Q., Huang, M., and Ma, X.: Statistical analysis of microphysical properties and the parameterization of effective radius of warm clouds in Beijing area, Atmos. Environ., 93, 888–896, 2009.

Dlugi, R.: Chemistry and Deposition of Soot Particles in Moist Air and Fog, Aerosol Science and Technology, 10, 93–105, 1989.

Doms, G. and Baldauf, M.: A Description of the Nonhydrostatic Regional COSMO-Model, Part I: Dynamics and Numerics, 2015.

Doms, G., Förstner, J., Heise, E., Herzog, H.-J., Mironov, D., Raschendorfer, M., Reinhard, T., Ritter, B., Schrodin, R., Schulz, J.-P., and Vogel, G.: A Description of the Nonhydrostatic Regional COSMO Model, Part II: Physical Parameterization, 2011.

Dung, E. J., Bombom, L. S., and Agusomu, T. D.: The effects of gas flaring on crops in the Niger Delta, Nigeria, GeoJournal, 73, 297–305, 2008.

EDGAR: Emission Database for Global Atmospheric Research, http://edgar.jrc.ec.europa.eu/htap_v2/index.php?SECURE=123 (last access: 10 June 2017), 2010.

Ekman, A. M. L.: Do sophisticated parameterizations of aerosol-cloud interactions in CMIP5 models improve the representation of recent observed temperature trends?, J. Geophys. Res., 119, 817–832, 2014.

Ekpoh, I. J. and Obia, A. E.: The role of gas flaring in the rapid corrosion of zinc roofs in the Niger Delta Region of Nigeria, The Environmentalist, 30, 347–352, 2010.

Elvidge, C. D., Zhizhin, M., Baugh, K., Hsu, F.-C., and Ghosh, T.: Methods for Global Survey of Natural Gas Flaring from Visible Infrared Imaging Radiometer Suite Data, Energies, 9, 2016.

EPA: AP 42, Fifth Edition, Compilation of Air Pollutant Emission Factors, Volume I: Stationary Point and Area Sources, Chapter 13: Miscellaneous Sources, 13.5 Industrial Flares Final Section, Environmental Protection Agency (https://www3.epa.gov/ttn/chief/ap42/ch13/final/C13S05_4-20-15.pdf, last access: 10 June 2017), 1, 1–6, 2015.

Fan, J., Leung, L. R., Rosenfeld, D., Chen, Q., Li, Z., Zhang, J., and Yan, H.: Microphysical effects determine macrophysical response for aerosol impacts on deep convective clouds, Proceedings of the National Academy of Sciences of the United States of America, 110, 4581–4590, 2013.

Fan, J., Wang, Y., Rosenfeld, D., and Liu, X.: Review of Aerosol-Cloud Interactions: Mechanisms, Significance, and Challenges, J. Atmos. Sci., 73, 4221–4250, 2016.

Fécan, F., Marticorena, B., and Bergametti, G.: Parametrization of the increase of the aeolian erosion threshold wind friction velocity due to soil moisture for arid and semi-arid areas, Ann. Geophysicae, 17, 149–157, 1999.

Fink, A. H., Engel, T., Ermert, V., van der Linden, R., Schneidewind, M., Redl, R., Afiesimama, E., Thiaw, W., Yorke, C., , and Evans, M.: Mean climate and seasonal cycle. In: Meteorology of tropical West Africa: The forecasters' handbook, John Wiley & Sons, Ltd, Chichester, UK., Parker, D. J. and Diop-Kane M. (Eds.) , pp. 1–39, 2017.

Flamant, C., Knippertz, P., Fink, A. H., Akpo, A., Brooks, B., Chiu, C., Coe, H., Danuor, S., Evans, M., Jegede, O., Kalthoff, N., Konaré, A., Liousse, C., Lohou, F., Mari, C., Schlager, H., Schwarzenboeck, A., Adler, B., Amekudzi, L., Aeyee, J., Ayoola, M., Bessardon, G., Bower, K., Burnet, F., Catoire, V., Colomb, A., Fossu-Amankwah, K., Lee, J., Lothon, M.,

Manaran, M., Marsham, J., Meynadier, R., Ngamini, J.-B., Rosenberg, P., Sauer, D., Schneider, J., Smith, V., Stratmann, G., Voigt, C., and Yoboue, V.: The Dynamics-Aerosol-Chemistry-Cloud Interactions in West Africa field campaigns: Overview and research highlights, Bull. Amer. Meteor. Soc., in press, 2017.

Fountoukis, C. and Nenes, A.: Continued development of a cloud droplet formation parameterization for global climate models, J. Geophys. Res., 110, 2005.

Fountoukis, C. and Nenes, A.: ISORROPIA II: a computationally efficient thermodynamic equilibrium model for K^+ -Ca^{2+} -Mg^{2+} -NH_4^+ -SO_4^{2-} -NO_3^- -CL^--H_2O aerosol, Atmos. Chem. Phys., 7, 4639–4659, 2007.

Freitas, S. R., Longo, K. M., and Andreae, M. O.: Impact of including the plume rise of vegetation fires in numerical simulations of associated atmospheric pollutants, Geophys. Res. Lett., 33, 2006.

Freitas, S. R., Longo, K. M., Chatfield, R., Latham, D., Silva Dias, M. A. F., Andreae, M. O., Prins, E., Santos, J. C., Gielow, R., and Carvalho Jr., J. A.: Including the sub-grid scale plume rise of vegetation fires in low resolution atmospheric transport models, Atmos. Chem. Phys., 7, 3385–3398, 2007.

Freitas, S. R., Longo, K. M., Trentmann, J., , and Latham, D.: Sensitivity of 1-D smoke plume rise models to the inclusion of environmental wind drag, Atmos. Chem. Phys., 10, 585–594, 2010.

Friedlander, S. K.: Smoke, Dust and Haze, John Wiley and Sons, New York, 1977.

Fuller, K. A., Malm, W. C., and Kreidenweis, S. M.: Effects of mixing on extinction by carbonaceous particles, J. Geophys. Res., 104, 15 941–15 954, 1999.

Gbambie, A. S. B. and Steyn, D. G.: Sea breezes at Cotonou and their interaction with the West African monsoon, Int. J. Climatol., 33, 2889–2899, 2013.

GlobCover: GlobCover Land Cover Map - European Space Agency GlobCover Project (http://www.gelib.com/globcover-2009.htm, last access: 10 June 2017), 2009.

GoogleEarth: Image © 2014 DigitalGlobe (https://www.google.com/intl/de/earth/, last access: 10 June 2017), 2014.

Grams, C. M., Jones, S. C., Marsham, J. H., Parker, D. J., Haywood, J. M., and Heuveline, V.: The Atlantic Inflow to the Saharan heat low: Observations and Modelling, Quart. J. Roy. Meteor. Soc., 136, 125–140, 2010.

Grant, L. D. and van den Heever, S. C.: Aerosol-cloud-land surface interactions within tropical sea breeze convection, J. Geophys. Res., 119, 8340–8361, 2014.

Grythe, H., Ström, J., Krejci, R., Quinn, P., and Stohl, A.: A review of sea-spray aerosol source functions using a large global set of sea salt aerosol concentration measurements, Atmos. Chem. Phys., 14, 1277–1297, 2014.

Guenther, A. B., Jiang, X., Heald, C. L., Sakulyanontvittaya, T., Duhl, T., Emmons, L. K., and Wang, X.: The Model of Emissions of Gases and Aerosols from Nature version 2.1 (MEGAN2.1): and extended and updated framework for modeling biogenic emissions, Geosci. Model Dev., 5, 1471–1492, 2012.

Guigard, S. E., Kindzierski, W. B., and Harper, N.: Heat Radiation from Flares. Report prepared for Science and Technology Branch, Alberta Environment, Alberta Environment, ISBN 0-7785-1188-X, Edmonton, Alberta , 2000.

Hannak, L., Knippertz, P., Fink, A. H., Kniffka, A., and Pante, G.: Why Do Global Climate Models Struggle to Represent Low-Level Clouds in the West African Summer Monsoon?, J. Climate, 30, 1665–1687, 2017.

Harris, G. W., Carter, P. L., Winer, A. M., and Pitts, J. N.: Observations of Nitrous Acid in the Los Angeles Atmosphere and Implications for Predictions of Ozone-Precursor Relationships, Environ. Sci. Technol., 16, 414–419, 1982.

Hayward, D. and Oguntoyinbo, J.: Climatology of West Africa, Rowman & Littlefield, p. 117, 1987.

Hill, P. G., Allan, R. P., Chiu, J. C., and Stein, T. H. M.: A multisatellite climatology of clouds, radiation, and precipitation in southern West Africa and comparison to climate models, J. Geophys. Res., 121, 10 857–10 879, 2016.

Hill, P. G., Allan, R. P., and Chiu, J. C.: Quantifying the contribution of different cloud types to the radiation budget, Atmos. Chem. Phys., under review, 2017.

Hopkins, J. R., Evans, M. J., Lee, J. D., Lewis, A. C., Marsham, J. H., McQuaid, J. B., Parker, D. J., Steward, D. J., Reeves, C. E., and Purvis, R. M.: Direct estimates of emissions from the megacity of Lagos, Atmos. Chem. Phys., 9, 8471–8477, 2009.

Hsu, N. C., Jeong, M.-J., Bettenhausen, C., Sayer, A. M., Hansell, R., Senftor, C. S., Huang, J., and Tsay, S.-C.: Enhanced Deep Blue aerosol retrieval algorithm: The second generation, J. Geophys. Res., 118, 9296–9315, 2013.

Huffman, G.: TRMM (TMPA-RT) Near Real-Time Precipitation L3 3 hour 0.25 degree x 0.25 degree V7, Greenbelt, MD, Goddard Earth Sciences Data and Information Services Center (GES DISC), https://disc.gsfc.nasa.gov/datacollection/TRMM_3B42RT_7.html (last access: 10 June 2017), 2016.

Hughes, L. S., Allen, J. O., Bhave, P., Kleeman, M. J., Cass, G. R., Liu, D.-Y., Fergenson, D. P., Morrical, B. D., and Prather, K. A.: Evolution of Atmospheric Particles along trajectories Crossing the Los Angeles Basin, Environ. Sci. Technol., 34, 3058–3068, 2000.

HWSD: HWSD (FAO/IIASA/ISRIC/ISSCAS/JRC), 2012: Harmonized World Soil Database (version 1.2), FAO, Rome, Italy and IIASA, Laxenburg, Austria, 2012.

IPCC: Climate Change 2013: The Physical Science Basis. Contribution of Working Group I to the Fifth Assessment Report of the Intergovernmental Panel on Climate Change; Stocker, T. F. and Qin, G. and Plattner, K. and Tignor, M. and Allen, S. K. and Boschung, J. and Nauels, A. and Xia, Y. and Bex, V. and Midgley, P. M. (eds.), Cambridge University Press, Cambridge, United Kingdom and New York, NY, USA, 8, 2013.

Ismail, O. S. and Umukoro, G. E.: Modelling combustion reactions for gas flaring and its resulting emissions, Journal of King Saud University – Engineering Sciences, 28, 130–140, 2014.

Jacobson, M. Z.: A physically-based treatment of elemental carbon optics: Implications for global direct forcing of aerosols, Geophys. Res. Lett., 27, 217–220, 2000.

Johnson, M. R., Devillers, R. W., and Thomson, K. A.: Quantitative Field Measurements of Soot Emission from Large Gas Flare using Sky-LOSA, Environ. Sci. Technol., 45, 345–350, 2011.

Kaiser, J. W., Heil, A., Andreae, M. O., Benedetti, A., Chubarova, N., Jones, L., Morcrette, J.-J., Razinger, M., Schultz, M. G., Suttie, M., and van der Werf, G. R.: Biomass burning emissions estimated with a global fire assimilation system based on observed fire radiative power, Biogeosciences, 9, 527–554, 2012.

Kalthoff, N., Adler, B., Wieser, A., Kohler, M., Träumner, K., Handwerker, J., Corsmeier, U., Khodayar, S., Lambert, D., Kopmann, A., Kunka, N., Dick, G., Ramatschi, M., Wickert, J., and Kottmeier, C.: KITcube - a mobile observation platform for convective studies deployed during HyMeX, Meteorologische Zeitschrift, 22, 633–647, 2013.

Kalthoff, N., Louhou, F., Brooks, B., Jegede, O., Adler, B., Babić, K., Dione, C., Ajao, A., Amekudzi, L. K., Aryee, J. N. A., Bessardon, G., Danuor, S. K., Handwerker, J., Kohler, M., Lothon, M., Pedruzo-Bagazgoitia, X., Smith, V., Sunmonu, L., Wieser, A., Fink, A. H., and Knippertz, P.: The diurnal cycle of the southern West African atmospheric boundary layer during the DACCIWA campaign, Atmos. Chem. Phys., under review, 2017.

Kerminen, V.-M. and Wexler, A. S.: Post-fog nucleation of H_2SO_4-H_20 particles in smog, Atmos. Environ., 28, 2399–2406, 1994.

Kettle, A. J., Andreae, M. O., Amouroux, D., Andreae, T. W., Bates, T. S., Berresheim, H., Bingemer, H., Boniforti, R., Curran, M. A. J., DiTullio, G. R., Helas, G., Jones, G. B., Keller, M. D., Kiene, R. P., Leck, C., Levasseur, M., Malin, G., Maspero, M., Matrai, P., McTaggart, A. R., Mihalopoulos, N., Nguyen, B. C., Novo, A., Putaud, J. P., Rapsomanikis, S., Roberts, G., Schebeske, G., Sharma, S., Simó, R., Staubes, R., Turner, S., and Uher, G.: A global database of sea surface dimethyl sulfide (DMS) measurements and a procedure to predict sea surface DMS as a function of latitude, longitude, and month, Global Biogeochem. Cycles, 13(2), 399–444, 1999.

Knippertz, P., Coe, H., Chiu, J. C., Evans, M. J., Fink, A. H., Kalthoff, N., Liousse, C., Mari, C., Allan, R. P., Brooks, B., Danour, S., Flamant, C., Jegede, O. O., Fabienne, L., and Marsham, J. H.: The DACCIWA Project, Bull. Amer. Meteor. Soc., pp. 1451–1460, 2015a.

Knippertz, P., Evans, M. J., Field, P. R., Fink, A. H., Liousse, C., and Marsham, J. H.: The possible role of local air pollution in climate change in West Africa, Nature Climate Change, 5, 814–822, 2015b.

Knippertz, P., Fink, A. H., Deroubaix, A., Morris, E., Tocquer, F., Evans, M. J., Flamant, C., Gaetani, M., Lavaysse, C., Mari, C., Marsham, J. H., Meynadier, R., Affo-Dogo, A., Bahaga, T., Brosse, F., Deetz, K., Guebsi, R., Latifou, I., Maranan, M., Rosenberg, P. D., and Schlüter, A.: A meteorological and chemical overview of the DACCIWA field campaign in West Africa in June-July 2016, Atmos. Chem. Phys., pp. 10 893–10 918, 2017.

Knote, C., Brunner, D., Vogel, H., Allan, J., Asmi, A., Äijälä, M., Carbone, S., van der Gon, H. D., Jimenez, J. L., Kiendler-Scharr, A., Mohr, C., Poulain, L., Prévôt, A. S. H., Swietlicke, E., and Vogel, B.: Towards an online-coupled chemistry-climate model: evaluation of trace gases and aerosols in COSMO-ART, Geosci. Model Dev., 4, 1077–1102, 2011.

Köhler, H.: The nucleus in and the growth of hygroscopic droplets, Trans. Faraday Soc., 32, 1152–1161, 1936.

Koren, I., Altaratz, O., Remer, L. A., Feingold, G., and Martins, J. V.: Aerosol-induced intensification of rain from the tropics to the mid-latitudes, Nature Geoscience, 5, 118–122, 2012.

Koren, I., Dagan, G., and Altaratz, O.: From aerosol-limited to invigoration of warm convective clouds, Science, 344, 1143–1146, 2014.

Kraut, I.: Separating the Aerosol Effect in Case of a "Medicane", Wissenschaftliche Berichte des Instituts für Meteorologie und Klimaforschung des Karlsruher Instituts für Technologie, Dissertation, 2015.

Kruger, S. K., McLean, G. T., and Fu, Q.: Numerical Simulation of the Stratus-to-Cumulus Transition in the Subtropical Marine Boundary Layer. Part I: Boundary-Layer Structure, J. Atmos. Sci., 52, 2839–2850, 2015.

Kumar, P., Sokolik, I. N., and Nenes, A.: Parameterization of cloud droplet formation for global and regional models: including adsoption activation from insoluble CCN, Atmos. Chem. Phys., 9, 2517–2532, 2009.

Lafore, J. P., Flamant, C., Guichard, F., Parker, D. J., Bouniol, D., Fink, A. H., Giraud, V., Gosset, M., Hall, N., Höller, H., Jones, S. C., Protat, A., Roca, R., Roux, F., Saïd, F., and Thomcroft, C.: Progress in understanding of weather systems in West Africa, Atmos. Sci. Lett., 12, 7–12, 2011.

Laing, A. and Evans, J.-L.: Introduction to Tropical Meteorology 2nd Edition, A Comprehensive Online & Print Textbook, University Corporation for Atmospheric Research, 2011.

Lana, A., Bell, T. G., Simó, R., Vallina, S. M., Ballabrera-Poy, J., Kettle, A. J., Dachs, J., Bopp, L., Saltzman, E. S., Stefels., J., Johnson, J. E., and Liss, P. S.: An updated climatology of surface dimethylsulfide concentrations and emission fluxes in the global ocean, Global Biogeochem. Cycles, 25, G1004, 2011.

Latham, J., Rasch, P., Chen, C.-C., Kettles, L., Gadian, A., Gettelman, A., Morrison, H., Bower, K., and Choularton, T.: Global temperature stabilization via controlled albedo enhancement of low-level maritime clouds, Phil. Trans. R. Soc., 366, 3969–3987, 2008.

Lau, K.-M. and Kim, K.-M.: Observational relationships between aerosol and Asian monsoon rainfall, and circulation, Geophys. Res. Lett., 33, 2006.

Lau, W. K. M., Kim, K.-M., Shi, J.-J., Matsui, T., Chin, M., Tan, Q., Peters-Lidard, C., and Tao, W. K.: Impacts of aerosol-monsoon interaction on rainfall and circulation over Northern India and the Himalaya Foothills, Climate Dynamics, 49, 1945–1960, 2017.

Lawrence, P. and Chase, T.: Representing a MODIS consistent land surface in the Community Land Model (CLM 3.0)., J. Geophys. Res., 112, G01 023, 2007.

Liousse, C., Assamoi, E., Criqui, P., Granier, C., and Rosset, R.: Explosive growth in African combustion emissions from 2005 to 2030, Environ. Res. Lett., 9, 1–10, 2014.

Lu, M., Conant, W. C., Jonsson, H. H., Varutbangkul, V., Flagan, R. C., and Seinfeld, J. H.: The Marine Stratus/Stratocumulus Experiment (MASE): Aerosol-cloud relationships in marine stratocumulus, J. Geophys. Res., 112, 2007.

Ludwig, J., Meixner, F. X., Vogel, B., and Förstner, J.: Soil-air exchange of nitric oxide: An overview of processes, environmental factors, and modeling studies, Biogeochemistry, 52, 225–257, 2001.

Lundgren, K.: Direct Radiative Effects of Sea Salt on the Regional Scale, Institut für Meteorologie und Klimaforschung, Karlsruher Institut für Technologie, Dissertation, 2010.

Lundgren, K., Vogel, B., Vogel, H., and Kottmeier, C.: Direct radiative effects of sea salt for the Mediterranean region under conditions of low to moderate wind speeds, J. Geophys. Res., 118, 1906–1923, 2013.

Mari, C. H., Cailley, G., Corre, L., Saunois, M., Attié, J. L., Thouret, V., and Stohl, A.: Tracing biomass burning plumes from the Southern Hemisphere during the AMMA 2006 wet season experiment, Atmos. Chem. Phys., 8, 3951–3961, 2008.

Mari, C. H., Reeves, C. E., Law, K. S., Ancellet, G., Andrés-Hernández, M. D., Barret, B., Bechera, J., Borbon, A., Bouarar, I., Cairo, F., Commane, R., Delon, C., Evans, M. J., Fierli, F., Floquet, C., Galy-Lacaux, C., Heard, D. E., Homan, C. D., Ingham, T., Larsen, N., Lewis, A. C., Liousse, C., Murphy, J. G., Orlandi, E., Oram, D. E., Saunois, M., Serça, D., Steward, D. J., Stone, D., Thouret, V., van Velthoven, P., and Williams, J. E.: Atmospheric composition of West Africa: highlights from the AMMA international program, Atmos. Sci. Lett., 12, 13–18, 2011.

Marsham, J. H., Dixon, N. S., Garcia-Carreras, L., Lister, G. M. S., Parker, D. J., Knippertz, P., and Birch, C. E.: The role of moist convection in the West African monsoon system: Insights from continental-scale convection-permitting simulations, Geophys. Res. Lett., 40, 1843–1849, 2013.

Marticorena, B., Chatenet, B., Rajot, J. L., Traoré, S., Coulibaly, M., Diallo, A., Koné, I., Maman, A., NDiaye, T., and Zakou, A.: Temporal variability of mineral dust concenetrations over West Africa: analyses of a pluriannual monitoring from the AMMA Sahelian Dust Transect, Atmos. Chem. Phys., 10, 8899–8915, 2010.

Martin, S. T., Andreae, M. O., Artaxo, P., Baumgardner, D., Chen, Q., Goldstein, A. H., Guenther, A., Heald, C. L., Mayol-Bracero, O. L., McMurry, P. H., Pauliquevis, T., Pöschl, U., Prather, K. A., Roberts, G. C., Saleska, S. R., Dias, M. A. S., Spracklen, D. V., Swietlicki, E., and Trebs, I.: Sources and Properties of Amazonian Aerosol Particles, Rev. Geophys., 48, 2010.

Mathon, V., Diedhiou, A., and Laurent, H.: Relationship between easterly waves and mesoscale convective systems over the Sahel, Geophys. Res. Lett., 29, 1–4, 2002a.

Mathon, V., Laurent, H., and Lebel, T.: Mesoscale Convective System Rainfall in the Sahel, J. Appl. Meteor., 41, 1081–1092, 2002b.

Matsuki, A., Quennehen, B., Schwarzenboeck, A., Crumeyrolle, S., Venzac, H., Laj, P., and Gomes, L.: Temporal and vertical variations of aerosol physical and chemical properties over West Africa: AMMA aircraft campaign in summer 2006, Atmos. Chem. Phys., 10, 8437–8451, 2010.

May, P. T.: The Australian nocturnal jet and diurnal variations of boundary-layer winds over Mt. Isa in north-eastern Australia, Quart. J. Roy. Meteor. Soc., 121, 987–1003, 1995.

McComiskey, A., Feingold, G., Frisch, A. S., Turner, D. D., Miller, M. A., Chiu, J. C., Min, Q., and Ogren, J. A.: An assessment of aerosol-cloud interactions in marine stratus clouds based on surface remote sensing, J. Geophys. Res., 114, 2009.

Miles, N. L., Verlinde, J., and Clothiaux, E. E.: Cloud Droplet Size Distributions in Low-Level Stratiform Clouds, J. Atmos. Sci., 57, 295–311, 2000.

Mircea, M. and Stefan, S.: A theoretical study of the microphysical parameterization of the scaventing coefficient as a function of precipitation type and rate, Atmos. Environ., 32, 2931–2938, 1998.

MODIS: MODIS Atmosphere Aerosol products 04_L2 (https://modis-atmos.gsfc.nasa.gov/products/aerosol, last access: 30 July 2017), 2017.

Monahan, E. C., Spiel, D. E., and Davidson, K. L.: A model of marine aerosol generation via whitecaps and wave disruption, Oceanic White-caps and Their Role in Air-Sea Exchange Processes, pp. 167–174, 1986.

Monks, P. S., Archibald, A. T., Colette, A., Cooper, O., Coyle, M., Derwent, R., Fowler, D., Granier, C., Law, K. S., Mills, G. E., Stevenson, D. S., Tarasowa, O., Thouet, V., von Schneidemesser, E., Sommariva, R., Wild, O., and Williams, M. S.: Troposphere ozone and its precursors from the urban to the global scale from air quality to short-lived climate forcer, Atmos. Chem. Phys., 15, 8889–8973, 2015.

Montgomery, D. C.: Design and Analysis of Experiments, 5th ed., John Wiley, New York, 115, 684–686, 2005.

Morales, R. and Nenes, A.: Characteristic updrafts for computing distribution-averaged cloud droplet number and stratocumulus cloud properties, J. Geophys. Res., 115, 2010.

MOZART: MOZART-4/GEOS-5 archive, National Center for Atmospheric Research (NCAR), University Corporation for Atmospheric Research, Atmospheric Chemistry Observations & Modeling (https://www.acom.ucar.edu/wrf-chem/mozart.shtml, last access: 30 July 2017), 2017a.

MOZART: MOZART-4/GEOS-5 forecasts, National Center for Atmospheric Research (NCAR), University Corporation for Atmospheric Research, Atmospheric Chemistry Observations & Modeling (https://www.acom.ucar.edu/acresp/AMADEUS/mz4_output/chemfcst/, last access: 30 July 2017), 2017b.

Mårtensson, E. M., Nilsson, E. D., de Leeuw, G., Cohen, L. H., and Hansson, H. C.: Laboratory simulations and parameterization of the primary marine aerosol production, J. Geophys. Res., 108, 2003.

Myhre, G., Shindell, D., Bréon, F.-M., Collins, W., Fuglestvedt, J., Huang, J., Koch, D., Lamarque, J.-F., Lee, D., Mendoza, B., Nakajima, T., Robock, A., Stephens, G., T., T., and H., Z.: Anthropogenic and Natural Radiative Forcing. In: Climate Change 2013: The Physical Science Basis. Contribution of Working Group I to the Fifth Assessment Report of the Intergovernmental Panel on Climate Change, Cambridge University Press, Cambridge, United Kingdom and New York, NY, USA, pp. 682–685, 694–698, 2013.

NASA: NASA Worldview (https://worldview.earthdata.nasa.gov/?p=geographic&l=MODIS_Terra_ CorrectedReflectance_TrueColor,Coastlines&t=2016-07-03&z=3&v=-14.440533542100408,-2.226047084313639, 15.090716457899592,13.910671665686362, last access: 17 August 2017), 2017.

Neubauer, D., Lohmann, U., Hoose, C., and Frontoso, M. G.: Impact of the representation of marine stratocumulus clouds on the anthropogenic aerosol effect, Atmos. Chem. Phys., 14, 11 997–12 022, 2014.

Nwankwo, C. N. and Ogagarue, D. O.: Effects of gas flaring on surface and ground waters in Delta State Nigeria, J. Geol. Min. Res., 3, 131–136, 2011.

Nwaugo, V. O., Onyeagba, R. A., and Nwahcukwu, N. C.: Effect of gas flaring on soil microbial spectrum in parts of Niger Delta area of southern Nigeria, Afr. J. Biotechnol., 5, 1824–1826, 2006.

Osuji, L. C. and Avwiri, G. O.: Flared Gas and Other Pollutants Associated with Air quality in industrial Areas of Nigeria: An Overview, Chemistry & Biodiversity, 2, 1277–1289, 2005.

Parker, D. J., Burton, R. R., Diongue-Niang, A., Ellis, R. J., Felton, M., M., T. C., Thorncroft, C. D., Bessemoulin, P., and Tompkins, A. M.: The diurnal cycle of the West African monsoon circulation, Quart. J. Roy. Meteor. Soc., 131, 2839–2860, 2005a.

Parker, D. J., Thorncroft, C. D., Burton, R. R., and Diongue-Niang, A.: Analysis of the African easterly jet, using aircraft observations from the JET2000 experiment, Quart. J. Roy. Meteor. Soc., 131, 1461–1482, 2005b.

Pearson, K. J., Lister, G. M. S., Birch, C. E., Allan, R. P., Hogan, R. J., and J., W. S.: Modeling the diurnal cycle of tropical convection across the 'grey zone', Quart. J. Roy. Meteor. Soc., 140, 491–499, 2013.

Peel, M. C., Finlayson, B. L., and A., M. T.: Updated world map of the Köppen-Geiger climate classification, Hydrol. Earth Syst. Sci., 11, 1633–1644, 2007.

Peng, J., Hu, M., Guo, S., Du, Z., Zheng, J., Shang, D., Zamora, M. L., Zeng, L., Shao, M., Wu, Y.-S., Zheng, J., Wang, Y., Glen, C. R., Collins, D. R., Molina, M. J., and Zhang, R.: Markedly enhanced absorption and direct radiative forcing of black carbon under polluted urban environments, PNAS, 113, 2016.

Philipps, V. T. J., DeMott, P. J., and Andronache, C.: An Empirical Parameterization of Heterogeneous Ice Nucleation for Multiple Chemical Species of Aerosol, J. Atmos. Sci., 65, 2757–2783, 2008.

R: R Core Team, R: A Language and Environment for Statistical Computing, R Foundation for Statistical Computing, Vienna, Austria, (http://www.R-project.org/, last access: 9 August 2017), 2013.

Raupach, M. R., Gillette, D. A., and Leys, J. F.: The effect of roughness elements on wind erosion threshold, J. Geophys. Res., 98, 3023–3029, 1993.

Real, E., Orlandi, E., Law, K. S., Fierli, F., Josset, D., Cairo, F., Schlager, H., Borrmann, S., Kunkel, D., Volk, C. M., McQuaid, J. B., Steward, D. J., Lee, J., Lewis, A. C., Hopkins, J. R., Ravegnani, F., Ulanovski, A., and Liousse, C.: Cross-hemispheric transport of central African biomass burning pollutants: implications for downwind ozone production, Atmos. Chem. Phys., 3027-3046, 2010.

Redelsperger, J.-L., Thorncroft, C. D., Diedhiou, A., Lebel, T., Parker, D. J., and Polcher, J.: African Monsoon Multidisciplinary Analysis - An International Research Project and Field Campaign, Bull. Amer. Meteor. Soc., pp. 1739–1746, 2006.

Rieger, D., Bangert, M., Kottmeier, C., Vogel, H., and Vogel, B.: Impact of aerosol on post-frontal convective clouds over Germany, Tellus, 66, 2014.

Rieger, D., Steiner, A., Bachmann, V., Gasch, P., Förstner, J., Deetz, K., Vogel, B., and Vogel, H.: Impact of a Saharan dust outbreak on the photovoltaic power generation in Germany, Atmos. Chem. Phys., 17, 13 391–13 415, 2017.

Riemer, N.: Numerische Simulationen zur Wirkung des Aerosols auf die troposphärische Chemie und die Sichtweite, Wissenschaftliche Berichte des Instituts für Meteorologie und Klimaforschung der Universität Karlsruhe, Dissertation, 2002.

Riemer, N., Vogel, H., and Vogel, B.: Soot aging time scales in polluted regions during day and night, Atmos. Chem. Phys., 4, 1885–1893, 2004.

Rinke, R.: Parametrisierung des Auswaschens von Aerosolpartikeln durch Niederschlag, Ph.D. thesis, Institut für Meteorologie und Klimaforschung der Universität Karlsruhe (TH), Karlsruhe, Germany, 2008.

Ritter, B. and Geleyn, J.-F.: A Comprehensive Radiation Scheme for Numerical Weather Prediction Models with Potential Application in Climate Simulations, Mon. Wea. Rev., 129, 303–325, 1992.

Rosenfeld, D.: TRMM Observed First Direct Evidence of Smoke from Forest Fires Inhibiting Rainfall, Geophys. Res. Lett., 26, 3105–3108, 1999.

Rosenfeld, D., Lohmann, U., Raga, G. B., O'Dowd, C., Kulmala, M., Fuzzi, S., Reissell, A., and Andreae, M. O.: Flood or Drought: How Do Aerosols Affect Precipitation?, Science, 321, 1309–1313, 2008.

Saleeby, S. M., Herbener, S. R., and van den Heever, S. C.: Impacts of Cloud Droplet-Nucleating Aerosols on Shallow Tropical Convection, J. Atmos. Sci., 72, 1369–1385, 2014.

Schrage, J. M. and Fink, A. H.: Nocturnal Continental Low-Level Stratus over Tropical West Africa: Observations and Possible Mechanisms Controlling Its Onset, Mon. Wea. Rev., 140, 1794–1809, 2012.

Schuster, R., Fink, A. H., and Knippertz, P.: Formation and Maintenance of Nocturnal Low-Level Stratus over the Southern West African Monsoon Region during AMMA 2006, J. Atmos. Sci., 70, 2337–2355, 2013.

Segal, Y. and Khain, A.: Dependence of droplet concentration on aerosol in different cloud types: Application to droplet concentration parameterization of aerosol conditions, J. Geophys. Res., 111, 2006.

Seifert, A. and Beheng, K. D.: A two-moment cloud microphysics parameterization for mixed-phase clouds. Part 1: Model description, Meteor. Atmos. Phys., 92, 45–66, 2006.

Seinfeld, J. H. and Pandis, S. N.: Atmospheric Chemistry and Physics, From Air Polution to Climate Change, John Wiley & Sons, Inc., Hoboken, New Jersey, 2, 2006a.

Seinfeld, J. H. and Pandis, S. N.: Atmospheric Chemistry and Physics, From Air Polution to Climate Change, John Wiley & Sons, Inc., Hoboken, New Jersey, 2, 795, 954, 2006b.

Shao, Y.: Simplification of a dust emission scheme and comparison with data, J. Geophys. Res., 109, D10 202, 2004.

Shao, Y. and Lu, H.: A simple expression for wind erosion threshold friction velocity, J. Geophys. Res., 105, 22 437–22 443, 2000.

Shao, Y., Fink, A., and Klose, M.: Numerical simulation of a continental-scale Saharan dust event, J. Geophys. Res., 115, D13 205, 2010.

Shao, Y., Wyrwoll, K.-H., Chappell, A., Huang, J., Lin, Z., McTainsh, G. H., Mikami, M., Tanaka, T. Y., Wang, X., and Yoon, S.: Dust cycle: An emerging core theme in Earth system science, Aeolian Research, 2, 181–204, 2011.

Sheffield, A. M., Saleeby, M., and van den Heever, S. C.: Aerosol-induced mechanisms for cumulus congestus growth, J. Geophys. Res., 120, 8941–8952, 2015.

Shrivastava, M., Berg, L. K., Fast, J. D., Easter, R. C., Laskin, A., Chapman, E. G., Gustafson, W. I., Liu, Y., and Berkowith, C. M.: Modeling aerosols and their interactions with shallow cumuli durng the 2007 CHAPS field study, J. Geophys. Res., 118, 1343–1360, 2013.

Simpson, J. E.: A comparison between laboratory and atmospheric density currents, Quart. J. Roy. Meteor. Soc., 95, 758–765, 1969.

Smith, M. H., Park, P. M., and Consterdine, I. E.: Marine aerosol concentrations and estimated fluxes over the ocean, Quart. J. Roy. Meteor. Soc., 119, 809–824, 1993.

Song, J., Liao, K., Coulter, R. L., and Lesht, B. M.: Climatology of the Low-Level Jet at the Southern Great Plains Atmospheric Boundary Layer Experiments Site, J. Appl. Meteor., 44, 1593–1606, 2005.

Sonibare, J. A. and Akeredolu, F. A.: A theoretical prediction of non-methane gaseous emissions from natural gas combustion, Energy Policy, 32, 1653–1665, 2004.

Sorjamaa, R. and Laaksonen, A.: The effect of H_20 adsorption on cloud drop activation of insoluble particles: a theoretical framework, Atmos. Chem. Phys., 7, 6175–6180, 2007.

Stanelle, T., Vogel, B., Vogel, H., Bäumer, D., and Kottmeier, C.: Feedback between dust particles and atmospheric processes over West Africa during dust episodes in March 2006 and June 2007, Atmos. Chem. Phys., 10, 10 771–10 788, 2010.

Stevens, B. and Feingold, G.: Untangling aerosol effects on clouds and precipitation in a buffered system, Nature Reviews, 461, 607–613, 2009.

Stockwell, W. R., Middleton, P., Chang, J. S., and Tang, X.: The second generation regional acid deposition model chemical mechanism for regional air quality modeling, J. Geophys. Res., 95, 16 343–16 367, 1990.

Stokes, R. H. and Robinson, R. A.: Interactions in aqueous nonelectrolyte solutions. I. Solute-solvent equilibria, J. Phys. Chem., 70, 2126–2130, 1966.

Strosher, M. T.: Investigations of flare gas emissions in Alberta, Final Report to: Environment Canada, Conservation and protection, the Alberta Energy and Utilities Board, and the Canadian Association of Petroleum Producers,

Environmental Technologies, Alberta Research Council, Calgary, Alberta, 1996.

Stull, R. B.: An Introduction to Boundary Layer Meteorology, Kluver Academic Publishers, Dortrecht, The Netherlands, pp. 520–526, 1988.

Sultan, B. and Janicot, S.: The West African Monsoon Dynamics. Part II: The "Preonset" and "Onset" of the Summer Monsoon, J. Climate, 16, 3407–3427, 2003.

Sun, J., Burns, S. P., Lenschow, D. H., Banta, R., Newsom, R., Coulter, R., Frasier, S., Ince, T., Nappo, C., Cuxart, J., Blumen, W., and Hu, X.-Z.: Intermittent turbulence associated with a density current passage in the stable boundary layer, Bound.-Layer Meteor., 105, 199–219, 2002.

Tao, W.-K., Lang, S., Simpson, J., Sui, C.-H., Ferrier, B., and Chou, M.-D.: Mechanisms of Cloud-Radiation Interaction in the Tropics and Midlatitudes, J. Atmos. Sci., 53, 2624–2651, 1996.

Tegen, I., Hollrig, P., Chin, M., Fung, I., Jacob, D., and J., P.: Contribution of different aerosol species to the global aerosol extinction optical thickness: Estimates from model results, J. Geophys. Res., 102, 23 895–23 915, 1997.

Teller, A. and Levin, Z.: Factorial method as a tool for estimating the relative contribution to precipitation of cloud microphysical processes and environmental conditions: Method and application, J. Geophys. Res., 113, 2008.

Tompkins, A. M., Cardinali, C., Morcrette, J.-J., and Rodwell, M.: Influence of aerosol climatology on forecasts of the African Easterly Jet, Geophys. Res. Lett., 32, 2005.

Twomey, S.: The Influence of Pollution on the Shortwave Albedo of Clouds, J. Atmos. Sci., 34, 1149–1152, 1977.

UNO: World Urbanization Prospects: The 2014 Revision, (ST/ESA/SER.A/366)., United Nations, Department of Economic and Social Affairs, Population Division, 2015a.

UNO: World Population Prospects 2015 – Data Booklet (ST/ESA/SER.A/377)., United Nations, Department of Economic and Social Affairs, 2015b.

UNO: The World's Cities in 2016 - Data Booklet (ST/ESA/SER.A/392)., United Nations, Department of Economic and Social Affairs, Population Division, 2016.

van de Wiel, B. J. H.: A Conceptual View on Inertial Oscillations and Nocturnal Low-Level Jets, J. Atmos. Sci., 67, 2679–2689, 2010.

van der Linden, R., Fink, A. H., and Redl, R.: Satellite-based climatology of low-level continental clouds in southern West Africa during the summer monsoon season, J. Geophys. Res., 120, 1186–1201, 2015.

VDI: Dispersion of Air Pollutants in the Atmosphere, Determination of plume rise (https://www.vdi.de/richtlinie/vdi_3782_blatt_3-ausbreitung_von_luftverunreinigungen_in_der_atmosphaere_berechnung_der_abgasfahnenueberhoehung/, last access: 10 June 2017), Verein Deutscher Ingenieure, VDI-Richtlinien, 3782, Part 3, Equation 24, 1985.

VEU: Transport and the Environment (VEU), http://www.dlr.de/VEU/en/desktopdefault.aspx/16635_read-41601/ (last access: 10 June 2017), 2017.

VIIRS: VIIRS Nighttime Detection and Characterization of Combustion Sources (http://ngdc.noaa.gov/eog/viirs/download_viirs_fire.html, last access: 10 June 2017), Defense Meteorological Satellite Program (DMSP), Earth Observation Group (EOG), 2016.

Vinken, G. C. M., Boersma, K. F., Maasakkers, J. D., Adon, M., and Martin, R. V.: Worldwide biogenic soil NOx emissions inferred from OMI NO2 observations, Atmos. Chem. Phys., 14, 10 363–10 381, 2014.

Vogel, B., Vogel, H., Kleffmann, J., and Kurtenbach, R.: Measured and simulated vertical profiles of nitrous acid - Part II. Model simulations and indications for a photolytic source, Atmos. Environ., 37, 2957–2966, 2003.

Vogel, B., Hoose, C., Vogel, H., and Kottmeier, C.: A model of dust transport applied to the Dead Sea Area, Meteorologische Zeitschrift, 15, 611–624, 2006.

Vogel, B., Vogel, H., Bangert, M., Lundgren, K., Rinke, R., and Stanelle, T.: The comprehensive model system COSMO-ART - Radiative impact of aerosol on the state of the atmosphere on the regional scale, Atmos. Chem. Phys., 9, 8661–8680, 2009.

von Hoyningen-Huene, W. and Posse, P.: Nonsphericity of Aerosol Particles and their Contribution to Radiative Forcing, Journal of Quantitative Spectroscopy & Radiative Transfer, 57, 651–668, 1997.

Walter, C., Freitas, S. R., Kottmeier, C., Kraut, I., Rieger, D., Vogel, H., and Vogel, B.: The importance of plume rise on the concentrations and atmospheric impacts of biomass burning aerosol, Atmos. Chem. Phys., 16, 9201–9219, 2016.

Wegener, A.: Thermodynamik der Atmosphäre, J. A. Barth, Leipzig, p. 311, 1911.

Weimer, M., Schröter, J., Eckstein, J., Deetz, K., Neumaier, M., Fischbeck, G., Hu, L., Millet, D. B., Rieger, D., Vogel, H., Vogel, B., Reddmann, T., Kirner, O., Ruhnke, R., and Braesicke, P.: An emission module for ICON-ART 2.0: implementation and simulations of acetone, Geosci. Model Dev., 10, 2471–2494, 2017.

Weingartner, E., Burtscher, H., and Baltensperger, U.: Hygroscopic properties of carbon and diesel soot particles, Atmos. Environ., 31, 2311–2327, 1997.

Whitby, E. R., McMurray, P. H., Shankar, U., and Binkowski, F. S.: Modal Aerosol Dynamics Modeling, Technical Report 600/3-91/020, (NTIS PB91-161729/AS Natl. Tech. Inf. Serv. Springfield, Atmos. Res. and Exposure Assess. Lab. U.S. Environ. Prot. Agency, Research Triangle Park, N.C., 1991.

Whitby, K. T.: The physical characteristics of sulfur aerosols, Atmos. Environ., 12, 135–159, 1978.

White, B., R.: Soil transport by winds on mars, J. Geophys. Res., 84, 4643–4651, 1979.

Xue, H. and Feingold, G.: Large-Eddy Simulations of Trade Wind Cumuli: Investigation of Aerosol Indirect Effects, J. Atmos. Sci., 63, 1605–1622, 2006.

Zdanovskii, A. B.: New methods for calculation solubilities of electrolytes in multicomponent systems, Zhur. Fiz. Kim., 22, 1475–1485, 1948.

Zhang, X. and Kondragunta, S.: Temporal and spatial variability in biomass burned areas across the USA derived from the GOES fire product, Remote Sens. Environ., 112, 2886–2897, 2008.

Zängl, G., Reinert, D., Rípidas, and Baldauf, M.: The ICON (ICOsahedral Non-hydrostatic) modelling framework of DWD and MPI-M: Description of the non-hydrostatic dynamical core, Quart. J. Roy. Meteor. Soc., 141, 563–579, 2015.

List of Figures

List of Tables

Acronyms

ACC	COSMO-ART aerosol mode: accumulation mode
ACCS	COSMO-ART aerosol mode: accumulation mode containing a soot core
ADE	Aerosol Direct Effect
AERONET	Aerosol Robotic Network
AEJ	African Easterly Jet
AEW	African Easterly Wave
AGL	Above Ground Level
AIE	Aerosol Indirect Effect
AIT	COSMO-ART aerosol mode: aitken mode
AITS	COSMO-ART aerosol mode: aitken mode containing a soot core
ALADINA	Unmanned Aerial System Carolo P360 Application of Light-weight Aircraft for Detecting in-situ Aerosol
ALWC	Aerosol Liquid Water Content
AMMA	African Monsoon Multidisciplinary Analysis
AOD	Aerosol Optical Depth
ASL	Above Sea Level
ASOP	Aircraft Special Observing Period
BAS Twin Otter	British Antarctic Survey Twin Otter (TO)
BCM	Billion cubic meter
BVOC	Biogenic Volatile Organic Compound
CAMS	Copernicus Atmosphere Monitoring Service

CANTHRO	COSMO-ART aerosol mode: coarse mode of anthropogenic origin
CCN	Cloud Condensation Nuclei
CDF	Cumulative Distribution Function
CDNC	Cloud Droplet Number Concentration
CM SAF	The Satellite Application Facility on Climate Monitoring
COSMO-ART	Consortium for Small-scale Modeling - Aerosol Reactive Trace gases
CPU	Central Processing Unit
CSEAS	COSMO-ART aerosol mode: coarse mode of marine origin
CSEASA	COSMO-ART aerosol mode: sea salt fine
CSEASB	COSMO-ART aerosol mode: sea salt medium
CSEASC	COSMO-ART aerosol mode: sea salt coarse
CSOIL	COSMO-ART aerosol mode: coarse mode of mineral origin
CSOILA	COSMO-ART aerosol mode: mineral dust fine
CSOILB	COSMO-ART aerosol mode: mineral dust medium
CSOILC	COSMO-ART aerosol mode: mineral dust coarse
D1	COSMO-ART Domain 1
D2	COSMO-ART Domain 2
D3	COSMO-ART Domain 3
DACCIWA	Dynamics-Aerosol-Chemistry-Cloud Interactions in West Africa
DJF	December, January, February
DWD	German Weather Service (Deutscher Wetterdienst)
ECCAD	Emissions of atmospheric Compounds and Compilation of Ancillary Data
ECDF	Empirical Cumulative Distribution Function
ECMWF	European Center for Medium range Weather Forecasting

EDGAR HTAP	Emissions Database for Global Atmospheric Research Hemispheric Transport of Air Pollution
EMFE	Evening Monsoon Flow Enhancement
FHH	Frenkel-Halsey-Hill
ForHLR	Forschungshochleistungsrechner (Phase) I
FRP	Fire Radiative Power
GEOS-5	Goddard Earth Observing System Model, Version 5
GFAS	Global Fire Assimilation System
GFEI	Gas Flaring Emission Inventory
GRAALS	General Radiative Algorithm Adapted to Linear-type Solutions radiation scheme
HSS	Heidke Skill Score
HWSD	Harmonized World Soil Database
ICON	Icosahedral Nonhydrostatic Model
IN	Ice Nuclei
IOP	Intensive Operation Period
IPCC AR5	International Panel on Climate Change Assessment Report 5
IU14	Ismail and Umukoro (2014)
ITCZ	InterTropical Convergence Zone
ITD	InterTropical Discontinuity
JJA	June, July, August
KIT	Karlsruhe Institute of Technology
LAI	Leaf Area Index
MACC	Monitoring Atmospheric Composition and Climate
MADE	Modal Aerosol Dynamics Model for Europe
MAM	March, April, May
MCS	Mesoscale Convective System
MODIS	MODerate-resolution Imaging Spectroradiometer
MOZART	Model for OZone And Related chemical Tracers

MSG/SEVIRI	Meteosat Second Generation/Spinning Enhanced Visible and Infrared Imager
NCAR	National Center for Atmospheric Research
NLLJ	Nocturnal Low-Level Jet
NLLS	Nocturnal Low-Level Stratus
OLR	Outgoing Longwave Radiation
OMMS	One-Moment Microphysics Scheme
PAPA	Parameterization of Photolysis Frequencies for Atmospheric Modeling
PBL	Planetary Boundary Layer
PSS	Peirce Skill Score
PFA	Peak Fire Activity
PFT	Plant Functional Types
PM10	Particulate Matter with diameters up to 10 μm
PM2.5	Particulate Matter with diameters up to 2.5 μm
POFD	Probability Of False Detection
RADM2	Regional Acid Deposition Model
RADMKA	Regional Acid Deposition Model Version Karlsruhe
RH	Relative Humidity
RMSE	Root Mean Square Error
SAL	Saharan Air Layer
SHL	Saharan Heat Low
SCC	Steinbuch Centre for Computing
SCT	Stratus-to-Cumulus Transition
SOA	Secondary Organic Aerosol
SON	September, October, November
SOOT	COSMO-ART aerosol mode: pure (fresh) soot mode
SPSD	Soil Particle Size Distribution
SST	Sea Surface Temperature
SWA	Southern West Africa
SYN	Synergistic effect of ADE and AIE

TKE	Turbulent Kinetic Energy
TMMS	Two-Moment Microphysics Scheme
TRMM	Tropical Rainfall Measuring Mission
TMPA-RT	TRMM Multi-Satellite Precipitation Analysis, the Near Real-Time
TS	Thread Score
UNO	United Nations Organization
UPS	Université Toulouse III - Paul Sabatier
USDA	United States Department of Agriculture
UTC	Universal Time Coordinated
VBS	Volatility Basic Set
VIIRS	Visible Infrared Imaging Radiometer Suite
VNF	VIIRS NightFire Nighttime Detection and Characterization of Combustion Sources
VNF_{flare}	VNF preprocessed by the Earth Observation Group of the National Oceanic and Atmospheric Administration (NOAA)
VOC	Volatile Organic Compound
VT	Vegetation Type
WAM	West African Monsoon

Symbols

A/d_D	Curvature term
$A_{FHH} = 2.5$	Compound specific empirical constant
$A_N = 0.0123$	Constant
a_w	Water activity
$a_z = 5$	Constant
B/d_D	Solute term
$B_{FHH} = 1.2$	Compound specific empirical constant
b_k	Extinction coefficient at spectral band *k*
$\tilde{b}_{k,l}$	Fitted constant
$\overline{Ca}_{0,AIT-AIT}$	Intramodal coagulation within AIT
$\overline{Ca}_{0,AIT-ACC}$	Intermodal coagulation between AIT and ACC
$\overline{Ca}_{0,AIT-AITS}$	Intermodal coagulation between AIT and AITS
$\overline{Ca}_{0,AIT-ACCS}$	Intermodal coagulation between AIT and ACCS
$\overline{Ca}_{0,AIT-SOOT}$	Intermodal coagulation between AIT and SOOT
c_{AIE}	Contribution from AIE
c_{ADE}	Contribution from ADE
c_b	Landuse class *bare soil*
c_p	Specific heat capacity of air
c_p^*	Mean specific heat capacity of gas flaring emissions
c_s	Landuse class *sparse vegetation*
c_{SYN}	Contribution from SYN
$c_{white} = 0.75$	Constant
d_D	Aqueous solution droplet diameter

$d_{D,c}$	Critical aqueous solution droplet diameter
d_p	Particle diameter
$d_{pg,l}$	Median diameter of mode l
d_1	Diurnal cycle of biomass burning heat flux
d_2	Diurnal cycle of biomass burning emissions
$E(x)$	Expected value of the gamma distribution
E_i	Gas flaring emission of species i
e_l	Particle binding energy of mode l
e_w	Vapor pressure
e_w^0	Saturation vapor pressure
F	Gas flaring emission flow rate
F_a	Shortwave actinic flux
F_{ADE}	Factor to scale the aerosol mass concentration within the aerosol direct effect (ADE)
F_{AIE}	Factor to scale the aerosol number concentration within the aerosol indirect effect (AIE)
$F_h(d_p)$	Saltation flux
F_i	Biogenic emission flux of compound i
F_{kin}	Kinetic energy flux
$F^{M_{0,AIT}}$	Reynolds stress tensor of the 0th moment of AIT
$F_{v,l}$	Dust emission flux of dust mode l
f	Fraction of total reaction energy that is radiated
f_{ACC}	Fraction of mode ACC
f_{ACCS}	Fraction of mode ACCS
f_{AIT}	Fraction of mode AIT
f_{AITS}	Fraction of mode AITS
f_b	Fraction of bare soil
f_c	Clay content
f_e	Fraction of daily biomass burning emissions that are emitted during PFA
f_l	Aerosol mass fraction of mode l

f_r	Correction function for roughness
$f_{SOOT(AIT)}$	Soot fraction of mode AIT
$f_{SOOT(ACC)}$	Soot fraction of mode ACC
f_ξ	Correction function for soil moisture
g	Gravitational acceleration
g_k	Asymmetry factor at spectral band k
$\tilde{g}_{k,l}$	Fitted constant
H	Radiant heat
H_f	Emitted heat flow
H_{max}	Heat flux upper limit
H_{min}	Heat flux lower limit
ΔH_v	Latent heat of condensation
$h_s = 250$ m	Specific height for the determination of the EMFE front
j_i	Photolysis frequency of species i
K	Turbulent diffusion coefficient
k	Shape parameter of the gamma distribution
M	Total aerosol mass density
$\frac{\partial \hat{M}_{0,AIT}}{\partial t}$	Prognostic equation of the 0th moment of AIT
M_f	Molar mass of the fuel gas
M_i	Number of moles of species i per air volume
M_k	K-th moment of the aerosol distribution
M_w	Molecular mass of water
m_i	Number of moles of species i per kg water
m'_i	Mass of fuel gas species i
m_l	Total wet aerosol mass
m'_{total}	Total mass of the fuel gas
N	Total aerosol number density
N_c	Cloud droplet number concentration
$\overline{Nu_0}$	Nucleation
$n(d_p)$	Size dependent aerosol number density

$n_s(d_p)$	Size dependent aerosol surface area density
$n_v(d_p)$	Size dependent aerosol volume density
n_{sol}	Amount of substance of solute
P	Pressure
$P_w(w')$	Subgrid scale vertical velocity gaussian PDF
$p(d_p)$	SPSD
$p_f(d_p)$	SPSD for soil without aggregates (fully dispersed)
p_f	Fuel gas pressure
p_l	Partitioning function for the kinetic energy of dust mode l
$p_m(d_p)$	SPSD for conserved soil aggregates (minimal dispersed)
q_c	Cloud water
q_g	Graupel
q_i	Cloud ice
q_r	Rain
q_s	Snow
q_v	Water vapor
R	Universal gas constant
R^2	Coefficient of determination
S	Total aerosol surface area density
SS_{AIE}	Sum of squares of AIE
SS_{ADE}	Sum of squares of ADE
SS_{SYN}	Sum of squares of SYN
SS_{TOT}	Sum of squares of the total effect
s	Saturation of air with respect to the liquid water phase
s^2	Variance
s_s	Supersaturation of air with respect to the liquid water phase
$s_{s,c}$	Critical supersaturation
$s_{s,e}$	Equilibrium supersaturation from Köhler theory
$s_{s,e,FHH}$	Equilibrium supersaturation from FHH theory
$s_{s,max}$	Maximum supersaturation

T	Temperature
T_a	Ambient temperature
T_s	Source temperature
t	Local solar time
t_0	Expected value of the gaussian distribution of vegetation type dependent diurnal cycle of heat flux limits, emissions and fire area
$u_\star$	Friction velocity
$u_{\star t}(d_p)$	Threshold friction velocity for real soil
$u_{\star t,min}$	Minimum threshold friction velocity
$u_{\star t0}(d_p)$	Threshold friction velocity for ideal soil
V	Total aerosol volume density
$\hat{\mathbf{v}}$	Mean wind speed
$\bar{v}(F)$	Postfrontal meridional wind speed
$\overline{v}_{sed,0,AIT}$	Sedimentation velocity of the 0th moment of AIT
$\overline{W}_{0,AIT}$	Wet deposition of the 0th moment of AIT
dW/dt	Condensation rate onto droplets
w'	Subgrid scale vertical velocity
α	Significance level
α_w	Water activity in the particle
$\gamma = 3 \cdot 10^{-4}$ kg s^{-2}	Constant
γ_d	Disaggregation factor
γ_i	Activity factor for biogenic emissions
δ	Availability of combustion air
$\varepsilon_{i,j}$	Biogenic emission factor of compound i and PFT j
η	Combustion efficiency
Θ	Scale parameter of the gamma distribution
Θ'	Number of adsorbed water molecules divided by number of molecules in a water layer
$\theta_s = 302$ K	Specific isentrope for the determination of the EMFE front

λ	Wavelength of radiation
ξ	Gravimetric soil moisture
$\xi'(f_c)$	Minimum gravimetric soil moisture dependent on clay content
π	Pi
ρ_a	Air density
$\bar{\rho}_a$	Mean air density
ρ_f	Fuel gas density
ρ_p	Particle density
ρ_w	Water density
σ	Absorption cross section
σ^*	Standard deviation of the gaussian distribution of vegetation-type dependent diurnal cycle of heat flux limits, emissions and fire area
$\sigma_{g,l}$	Geometric standard deviation of mode l
σ_i	Surface tension at the interface between particle and gas
σ_w	Surface tension of water
τ	AOD
$\Phi(\lambda,T,P)$	Quantum yield of the photodissociation products
χ_j	Fractional coverage of PFT j
ω_1	Weighting 1 for gaussian distribution of vegetation-type dependent diurnal cycle of heat flux limits, emissions and fire area
ω_2	Weighting 2 for gaussian distribution of vegetation-type dependent diurnal cycle of heat flux limits, emissions and fire area
$\tilde{\omega}_{k,AIT}$	Fitted constant at spectral band k
$\tilde{\omega}_{k,ACC}$	Fitted constant at spectral band k
$\tilde{\omega}_{k,l}$	Fitted constant at spectral band k
$\tilde{\omega}_{k,SOOT}$	Fitted constant at spectral band k
BC	Black carbon

Ca^{2+}	Calcium ion
Cl^-	Chloride ion
CO	Carbon monoxide
CO_2	Carbon dioxide
H_2O	Water
H_2SO_4	Sulfuric acid
HNO_3	Nitric acid
HONO	Nitrous acid
K^+	Potassium ion
Mg^{2+}	Magnesium ion
Na^+	Natrium ion
NH_4^+	Ammonium
N_2O_5	Dinitrogen pentoxide
NO	Nitrogen oxide
NO_2	Nitrogen dioxide
NO_x	Sum of nitrogen oxide and nitrogen dioxide
NO_3^-	Nitrate
O_3	Ozone
OH^-	Hydroxyl radical
SO_2	Sulfur dioxide
SO_4^{2-}	Sulfate

Acknowledgments

The research leading to these results has received funding from the European union 7th Framework Programme (FP7/2007-2013) under Grant Agreement no. 603502 (EU project DACCIWA: Dynamics-aerosol-chemistry-cloud interactions in West Africa). Thanks a lot to the DACCIWA community for providing the comprehensive observational dataset that is highly valuable not only but especially for the modelers. I thank the principal investigators Andreas Wieser, Rachel T. Pinker, Brent Holben and Richard Damoah for their efforts in establishing and maintaining the AERONET sites KITcube_Save, Ilorin and Koforidua_ANUC. The visualizations and statistical data analyses of this study were realized by using R (2013). I also want to thank the SCC for providing me with resources of the Forschungshochleistungsrechner ForHLR (Phase) I, that allows to conduct the COSMO-ART aerosol/chemistry forecasts for the DACCIWA measurement campaign as well as the detailed case study simulations. In this respect I also thank the DWD for providing me with the ICON forecasts as the basis for the COSMO-ART forecasts.
Furthermore, I am grateful for the financial support I got from the Karlsruhe House of Young Scientists (KHYS) at KIT, that allows me to realize a two-week research stay at the University of Leeds, and from the Graduate School for Climate and Environment (GRACE), that provided me with financial support regarding a one-month research stay at the University of Manchester as well as the visit of professional trainings and a summer school. Of course I also thank John Marsham for hosting me at the University of Leeds and Gordon Mcfiggans and Hue Coe for hosting me at the University of Manchester.

I would like to express my gratitude to Prof. Christoph Kottmeier and Prof. Peter Knippertz for supervising my dissertation. Their remarks significantly improved this study. Thanks to Heike Vogel, Bernhard Vogel and Tobias Schad for supporting me in scientific concerns and sharing academic, scientific and technical experience. I really appreciated the cooperation with the WP 1 measurement folks. Thank you Norbert, Bianca and Karmen for the fruitful discussions.

Wissenschaftliche Berichte des Instituts für Meteorologie und Klimaforschung des Karlsruher Instituts für Technologie (0179-5619)

Bisher erschienen:

Nr. 1: *Fiedler, F. / Prenosil, T.*
Das MESOKLIP-Experiment. (Mesoskaliges Klimaprogramm im Oberrheintal). August 1980

Nr. 2: *Tangermann-Dlugi, G.*
Numerische Simulationen atmosphärischer Grenzschichtströmungen über langgestreckten mesoskaligen Hügelketten bei neutraler thermischer Schichtung. August 1982

Nr. 3: *Witte, N.*
Ein numerisches Modell des Wärmehaushalts fließender Gewässer unter Berücksichtigung thermischer Eingriffe. Dezember 1982

Nr. 4: *Fiedler, F. / Höschele, K. (Hrsg.)*
Prof. Dr. Max Diem zum 70. Geburtstag. Februar 1983 (vergriffen)

Nr. 5: *Adrian, G.*
Ein Initialisierungsverfahren für numerische mesoskalige Strömungs modelle. Juli 1985

Nr. 6: *Dorwarth, G.*
Numerische Berechnung des Druckiderstandes typischer Geländeformen. Januar 1986

Nr. 7: *Vogel, B.; Adrian, G. / Fiedler, F.*
MESOKLIP-Analysen der meteorologischen Beobachtungen von mesoskaligen Phänomenen im Oberrheingraben. November 1987

Nr. 8: *Hugelmann, C.-P.*
Differenzenverfahren zur Behandlung der Advektion. Februar 1988

Nr. 9: *Hafner, T.*
Experimentelle Untersuchung zum Druckwiderstand der Alpen. April 1988

Nr. 10: *Corsmeier, U.*
Analyse turbulenter Bewegungsvorgänge in der maritimen atmosphärischen Grenzschicht. Mai 1988

Nr. 11: *Walk, O. / Wieringa, J.(eds)*
Tsumeb Studies of the Tropical Boundary-Layer Climate. Juli 1988

Nr. 12: *Degrazia, G. A.*
Anwendung von Ähnlichkeitsverfahren auf die turbulente Diffusion in der konvektiven und stabilen Grenzschicht. Januar 1989

Nr. 13: *Schädler, G.*
Numerische Simulationen zur Wechselwirkung zwischen Landoberflächen und atmophärischer Grenzschicht. November 1990

Nr. 14: *Heldt, K.*
Untersuchungen zur Überströmung eines mikroskaligen Hindernisses in der Atmosphäre. Juli 1991

Nr. 15: *Vogel, H.*
Verteilungen reaktiver Luftbeimengungen im Lee einer Stadt – Numerische Untersuchungen der relevanten Prozesse. Juli 1991

Nr. 16: *Höschele, K.(ed.)*
Planning Applications of Urban and Building Climatology – Proceedings of the IFHP / CIB-Symposium Berlin, October 14-15, 1991. März 1992

Nr. 17: *Frank, H. P.*
Grenzschichtstruktur in Fronten. März 1992

Nr. 18: *Müller, A.*
Parallelisierung numerischer Verfahren zur Beschreibung von Ausbreitungs- und chemischen Umwandlungsprozessen in der atmosphärischen Grenzschicht. Februar 1996

Nr. 19: *Lenz, C.-J.*
Energieumsetzungen an der Erdoberfläche in gegliedertem Gelände. Juni 1996

Nr. 20: *Schwartz, A.*
Numerische Simulationen zur Massenbilanz chemisch reaktiver Substanzen im mesoskaligen Bereich. November 1996

Nr. 21: *Beheng, K. D.*
Professor Dr. Franz Fiedler zum 60. Geburtstag. Januar 1998

Nr. 22: *Niemann, V.*
Numerische Simulation turbulenter Scherströmungen mit einem Kaskadenmodell. April 1998

Nr. 23: *Koßmann, M.*
Einfluß orographisch induzierter Transportprozesse auf die Struktur der atmosphärischen Grenzschicht und die Verteilung von Spurengasen. April 1998

Nr. 24: *Baldauf, M.*
Die effektive Rauhigkeit über komplexem Gelände – Ein Störungstheoretischer Ansatz. Juni 1998

Nr. 25: *Noppel, H.*
Untersuchung des vertikalen Wärmetransports durch die Hangwindzirkulation auf regionaler Skala. Dezember 1999

Nr. 26: *Kuntze, K.*
Vertikaler Austausch und chemische Umwandlung von Spurenstoffen über topographisch gegliedertem Gelände. Oktober 2001

Nr. 27: *Wilms-Grabe, W.*
Vierdimensionale Datenassimilation als Methode zur Kopplung zweier verschiedenskaliger meteorologischer Modellsysteme. Oktober 2001

Nr. 28: *Grabe, F.*
Simulation der Wechselwirkung zwischen Atmosphäre, Vegetation und Erdoberfläche bei Verwendung unterschiedlicher Parametrisierungsansätze. Januar 2002

Nr. 29: *Riemer, N.*
Numerische Simulationen zur Wirkung des Aerosols auf die troposphärische Chemie und die Sichtweite. Mai 2002

Nr. 30: *Braun, F. J.*
Mesoskalige Modellierung der Bodenhydrologie.
Dezember 2002

Nr. 31: *Kunz, M.*
Simulation von Starkniederschlägen mit langer Andauer über Mittelgebirgen. März 2003

Nr. 32: *Bäumer, D.*
Transport und chemische Umwandlung von Luftschadstoffen im Nahbereich von Autobahnen – numerische Simulationen.
Juni 2003

Nr. 33: *Barthlott, C.*
Kohärente Wirbelstrukturen in der atmosphärischen Grenzschicht. Juni 2003

Nr. 34: *Wieser, A.*
Messung turbulenter Spurengasflüsse vom Flugzeug aus.
Januar 2005

Nr. 35: *Blahak, U.*
Analyse des Extinktionseffektes bei Niederschlagsmessungen mit einem C-Band Radar anhand von Simulation und Messung.
Februar 2005

Nr. 36: *Bertram, I.*
Bestimmung der Wasser- und Eismasse hochreichender konvektiver Wolken anhand von Radardaten, Modellergebnissen und konzeptioneller Betrachtungen. Mai 2005

Nr. 37: *Schmoeckel, J.*
Orographischer Einfluss auf die Strömung abgeleitet aus Sturmschäden im Schwarzwald während des Orkans „Lothar". Mai 2006

Nr. 38: *Schmitt, C.*
Interannual Variability in Antarctic Sea Ice Motion: Interannuelle Variabilität antarktischer Meereis-Drift.
Mai 2006

Nr. 39: *Hasel, M.*
Strukturmerkmale und Modelldarstellung der Konvektion über Mittelgebirgen. Juli 2006

Ab Band 40 erscheinen die Wissenschaftlichen Berichte des Instituts für Meteorologie und Klimaforschung bei KIT Scientific Publishing (ISSN 0179-5619). Die Bände sind unter www.ksp.kit.edu als PDF frei verfügbar oder als Druckausgabe bestellbar.

Nr. 40: *Lux, R.*
Modellsimulationen zur Strömungsverstärkung von orographischen Grundstrukturen bei Sturmsituationen
ISBN 978-3-86644-140-8

Nr. 41: *Straub, W.*
Der Einfluss von Gebirgswellen auf die Initiierung und Entwicklung konvektiver Wolken
ISBN 978-3-86644-226-9

Nr. 42: *Meißner, C.*
High-resolution sensitivity studies with the regional climate model COSMO-CLM
ISBN 978-3-86644-228-3

Nr. 43: *Höpfner, M.*
Charakterisierung polarer stratosphärischer Wolken mittels hochauflösender Infrarotspektroskopie
ISBN 978-3-86644-294-8

Nr. 44: *Rings, J.*
Monitoring the water content evolution of dikes
ISBN 978-3-86644-321-1

Nr. 45: *Riemer, M.*
Außertropische Umwandlung tropischer Wirbelstürme: Einfluss auf das Strömungsmuster in den mittleren Breiten
ISBN 978-3-86644-766-0

Nr. 46: *Anwender, D.*
Extratropical Transition in the Ensemble Prediction System of the ECMWF: Case Studies and Experiments
ISBN 978-3-86644-767-7

Nr. 47: *Rinke, R.*
Parametrisierung des Auswaschens von Aerosolpartikeln durch Niederschlag
ISBN 978-3-86644-768-4

Nr. 48: *Stanelle, T.*
Wechselwirkungen von Mineralstaubpartikeln mit thermo-dynamischen und dynamischen Prozessen in der Atmosphäre über Westafrika
ISBN 978-3-86644-769-1

Nr. 49: *Peters, T.*
Ableitung einer Beziehung zwischen der Radarreflektivität, der Niederschlagsrate und weiteren aus Radardaten abgeleiteten Parametern unter Verwendung von Methoden der multivariaten Statistik
ISBN 978-3-86644-323-5

Nr. 50: *Khodayar Pardo, S.*
High-resolution analysis of the initiation of deep convection forced by boundary-layer processes
ISBN 978-3-86644-770-7

Nr. 51: *Träumner, K.*
Einmischprozesse am Oberrand der konvektiven atmosphärischen Grenzschicht
ISBN 978-3-86644-771-4

Nr. 52: *Schwendike, J.*
Convection in an African Easterly Wave over West Africa and the Eastern Atlantic : A Model Case Study of Hurricane Helene (2006) and its Interaction with the Saharan Air Layer
ISBN 978-3-86644-772-1

Nr. 53: *Lundgren, K.*
Direct Radiative Effects of Sea Salt on the Regional Scale
ISBN 978-3-86644-773-8

Nr. 54: *Sasse, R.*
Analyse des regionalen atmosphärischen Wasserhaushalts unter Verwendung von COSMO-Simulationen und GPS-Beobachtungen
ISBN 978-3-86644-774-5

Nr. 55: *Grenzhäuser, J.*
Entwicklung neuartiger Mess- und Auswertungs-strategien für ein scannendes Wolkenradar und deren Anwendungsbereiche
ISBN 978-3-86644-775-2

Nr. 56: *Grams, C.*
Quantification of the downstream impact of extratropical transition for Typhoon Jangmi and other case studies
ISBN 978-3-86644-776-9

Nr. 57: *Keller, J.*
Diagnosing the Downstream Impact of Extratropical Transition Using Multimodel Operational Ensemble Prediction Systems
ISBN 978-3-86644-984-8

Nr. 58: *Mohr, S.*
Änderung des Gewitter- und Hagelpotentials im Klimawandel
ISBN 978-3-86644-994-7

Nr. 59: *Puskeiler, M.*
Radarbasierte Analyse der Hagelgefährdung in Deutschland
ISBN 978-3-7315-0028-5

Nr. 60: *Zeng, Y.*
Efficient Radar Forward Operator for Operational Data Assimilation within the COSMO-model
ISBN 978-3-7315-0128-2

Nr. 61: *Bangert, M. J.*
Interaction of Aerosol, Clouds, and Radiation on the Regional Scale
ISBN 978-3-7315-0123-7

Nr. 62: *Jerger, D.*
Radar Forward Operator for Verification of Cloud Resolving Simulations within the COSMO Model
ISBN 978-3-7315-0172-5

Nr. 63: *Maurer, V.*
Vorhersagbarkeit konvektiver Niederschläge : Hochauflösende Ensemblesimulationen für Westafrika
ISBN 978-3-7315-0189-3

Nr. 64: *Stawiarski, C.*
Optimizing Dual-Doppler Lidar Measurements of Surface Layer Coherent Structures with Large-Eddy Simulations
ISBN 978-3-7315-0197-8

Die Bände sind unter www.ksp.kit.edu als PDF frei verfügbar oder als Druckausgabe bestellbar.

Nr. 65: *Mahlke, H.*
Mechanismen der Auslösung hochreichender Konvektion im südwestdeutschen Mittelgebirgsraum
ISBN 978-3-7315-0203-6

Nr. 66: *Fosser, G.*
Precipitation statistics from regional climate model at resolutions relevant for soil erosion
ISBN 978-3-7315-0227-2

Nr. 67: *Adler, B.*
Boundary-Layer Processes Producing Mesoscale Water-Vapour Variability over a Mountainous Island
ISBN 978-3-7315-0247-0

Nr. 68: *Kraut, I.*
Separating the Aerosol Effect in Case of a „Medicane"
ISBN 978-3-7315-0405-4

Nr. 69: *Breil, M.*
Einfluss der Boden-Vegetation-Atmosphären Wechselwirkungen auf die dekadische Vorhersagbarkeit des Westafrikanischen Monsuns
ISBN 978-3-7315-0420-7

Nr. 70: *Lott, F. F.*
Wind Systems in the Dead Sea and Footprints in Seismic Records
ISBN 978-3-7315-0596-9

Nr. 71: *Rieger, D.*
Der Einfluss von natürlichem Aerosol auf Wolken über Mitteleuropa
ISBN 978-3-7315-0672-0

Nr. 72: *Loewe, K.*
Arctic mixed-phase clouds. Macro- and microphysical insights with a numerical model
ISBN 978-3-7315-0686-7

Nr. 73: *Piper, D. A.*
Untersuchung der Gewitteraktivität und der relevanten großräumigen Steuerungsmechanismen über Mittel- und Westeuropa
ISBN 978-3-7315-0701-7

Nr. 74: *Metzger, J.*
Wind Systems and Energy Balance in the Dead Sea Valley
ISBN 978-3-7315-0699-7

Nr. 75: *Deetz, K.*
Assessing the Aerosol Impact on Southern West African Clouds and Atmospheric Dynamics
ISBN 978-3-7315-0744-4